Christian Kühnert

Data-driven Methods for Fault Localization in Process Technology

Karlsruher Schriften zur Anthropomatik
Band 15
Herausgeber: Prof. Dr.-Ing. Jürgen Beyerer

Eine Übersicht über alle bisher in dieser Schriftenreihe
erschienenen Bände finden Sie am Ende des Buchs.

Data-driven Methods for Fault Localization in Process Technology

by
Christian Kühnert

Dissertation, Karlsruher Institut für Technologie (KIT)
Fakultät für Informatik, 2013

Impressum

Scientific
Publishing

Karlsruher Institut für Technologie (KIT)
KIT Scientific Publishing
Straße am Forum 2
D-76131 Karlsruhe

KIT Scientific Publishing is a registered trademark of Karlsruhe
Institute of Technology. Reprint using the book cover is not allowed.

www.ksp.kit.edu

Print on Demand 2013

ISSN 1863-6489
ISBN 978-3-7315-0098-8

Data-driven Methods for Fault Localization in Process Technology

zur Erlangung des akademischen Grades eines

Doktors der Ingenieurwissenschaften

von der Fakultät für Informatik
des Karlsruher Instituts für Technologie (KIT)

genehmigte

Dissertation

von

Christian Felix Kühnert

aus Lindenfels

Tag der mündlichen Prüfung: 15. Juli 2013

Erster Gutachter: Prof. Dr.-Ing. Jürgen Beyerer

Zweiter Gutachter: Prof. Dr.-Ing. Thomas Längle

Acknowledgements

The present work was accomplished while being a research assistant at the Fraunhofer Insitute of Optronics, System Technologies and Image Exploitation (IOSB). Over the past years this work has benefited from several people and my gratitude is especially to those mentioned below.

At first, I would like to give special thanks to Prof. Dr.-Ing. Jürgen Beyerer and Prof. Dr.-Ing. Thomas Längle. Thanks for all the interesting discussions, valuable advices and for giving me the opportunity to pursue my own ideas throughout the thesis.

The researchers from the department of 'Mess- Regel- und Diagnosesysteme' made my time rewarding and lots of fun, especially because of showing interest for each other's research. Most of all I must thank Thomas Stephan for discussions on his whiteboard, Peter Frühberger for keeping me on track while writing, Mathias Ziebarth for his endless patience and keeping me in physical shape, Christian Frey for the hardware, Oliver Krol and Thomas Bernard for their wealth of experience, Philipp Woock for the Latex support and Janya-Anurak Chettapong for showing me the Thai way of living.

I very much appreciated the support from Dr. Michael Heizmann, Dr. Lutz Gröll and Prof. Dr.-Ing. Ralf Mikut. Discussions with them yielded a good portion of this work.

Living in Karlsruhe would have been half enjoyable without meeting Ching-hsiu Huang and Felix Sawo. Thanks for all the fun and telling me the latest stock market prices.

I would also like to take the opportunity to thank my parents Rolf, Heide and my sister Katrin with her family who were persistently asking me if I do my thesis as an excuse to reduce my physical working capacity. Without them I would not be who I am today. Moreover, special thanks go to my mountaineering family in the French Alps who gave me the opportunity to spend many holidays with them.

Finally, I would like to thank my wife Claire who accepted that I had a bad short-term memory for things she told me while writing this thesis and for reminding me that there exists a non-engineering life.

Christian Kühnert

Zusammenfassung

Verfahrenstechnische Prozesse besitzen in der Regel eine große Anzahl an Prozessgrößen. Bei der Detektion von abnormalem Prozessverhalten, beispielsweise durch Überschreiten eines zuvor definierten Schwellwertes oder durch multivariate Fehlerdetektion, können diese Prozessgrößen einen Alarm generieren. Tritt eine Störung in einem zentralen Anlagenteil auf, kann dies zu einer Alarmflut führen, da die einzelnen Prozessgrößen miteinander verbundenen sind und die Störung durch die Anlage propagiert. Anlagenführer müssen dann unter Zeitdruck die wichtigsten Alarme selektieren, um die tatsächliche Fehlerursache detektieren zu können und wichtige Entscheidungen müssen möglicherweise unter einem erhöhten Stresslevel getroffen werden. Daher ist es nützlich, automatisiert die Prozessgrößen zu detektieren, welche sich am nächsten an der Ursache der Störung befinden.

Der in dieser Arbeit untersuchte Ansatz zur Fehlerlokalisierung sich anlagenweit auswirkender Fehler besteht darin, statistische Relationen und Zeitverschiebungen in den Messdaten der Prozessgrößen zu nutzen (bspw. durch Dynamiken in Übertragungssystemen oder Totzeiten), um die Ursache der Störung ausfindig zu machen und den Störungspropagationspfad rückzurechnen. Methodisch bedeutet dies, dass zwischen Prozessgrößen Ursache/Wirkzusammenhänge berechnet werden müssen. Prozessgrößen, welche einen großen kausalen Einfluss auf weitere Größen haben, kommen hierbei als Ursache der Störung in Betracht.

Für das automatisierte Erkennen kausaler Abhängigkeiten wurden dazu Verfahren entwickelt, welche auf Kreuzkorrelation, Granger-Kausalität, Transfer Entropie und Support Vector Machines basieren. Speziell für Support Vector Machines wurde ein neuer Ansatz aufbauend auf einer Variablenselektion zeitverschobener Eingangssignale entwickelt, um kausale Abhängigkeiten in Messdaten zu erkennen. Bei allen Methoden wurde darauf geachtet, dass sie zum Erkennen der Zusammenhänge aus einem qualitativ statistischen Teil und einem quantitativen Teil bestehen. Dies bedeutet, dass zunächst ein Signifikanztest durchgeführt wird, welcher überprüft, ob eine kausale Abhängigkeit zwischen den Prozessvariablen vorliegt.

Für den quantitativen Teil wurde in dieser Arbeit das Konzept der *kausalen Stärke* entwickelt, um den resultierenden Einfluss der Prozessgrößen aufeinander zu berechnen. Bei der kausalen Stärke handelt es sich um ein Maß, welches die Ursache/Wirkungsbeziehungen zwischen verschiedenen Prozessgrößen quantifiziert. Die Grundidee hierbei ist, dass bei einem statistisch signifikanten Zusammenhang zwischen zwei Prozessgrößen die Stärke des kausalen Zusammenhangs bei allen Verfahren durch einen Wert zwischen 0 (keine Ursache/Wirkung) und 1 (sicher erkannte Ursache/Wirkung) beschrieben wird. Die Auswertung jeder Methode resultiert schließlich in einer Kausalmatrix, welche die Zusammenhänge der gesamten verwendeten Prozessgrößen beschreibt und den resultierenden Fehlerpropagationspfad beinhaltet. Durch die Normierung auf Werte zwischen 0 und 1 ist es möglich, die resultierenden Matrizen eines jeden Verfahrens zu fusionieren und so robustere Ergebnisse bezüglich der Fehlerlokalisierung zu erzielen.

Da jede Methode ihre speziellen Eigenschaften besitzt, wurden zunächst Benchmark-Datensätze erstellt und die entwickelten Methoden getestet, um eine Bewertung der Vor- und Nachteile vorzunehmen. Die Benchmark-Datensätze beinhalten hierbei zum Beispiel lange Totzeiten, Nichtlinearitäten, Wechsel im Stationärwert oder Datensätze, in denen mehr als eine Störung auftritt. Als Praxistest wurden die entwickelten Methoden sowohl an einem Laborprozess als auch an einem industriellen Glasziehprozess untersucht. Beim Laborprozess handelt es sich um einen Aufbau, bei dem Wasser in zwei Behältern umgepumpt wird. Zum Testen von Fehlern wurden hierbei mehrere Versuche durchgeführt. Hierzu zählt beispielsweise eine Störung der Stromzuführung der Pumpe, ein Klemmen des Ventils und als externe Störung das Verstopfen eines Zuleitungsrohres. In allen Fällen war es möglich, durch die Fusion der Methoden die Fehlerursache zu lokalisieren. Beim Glasziehprozess wurden die Methoden dazu verwendet, den Fehlerpropagationspfad eines Defektes im

Glaszylinder, welcher in der Produktion von Glasstäben auftritt, zurückzuverfolgen.

Des Weiteren wurden verschiedene Methoden zur Visualisierung der Kausalmatrizen entwickelt, um eine einfache Analyse der Ergebnisse zu ermöglichen. Hierzu zählt der Vergleich der Methoden am gleichen Datensatz in Form eines Bar-Charts, die resultierenden Fehlerpropagationspfade durch einen partiell gerichteten Graphen sowie das Erzeugen einer Fehlerhitliste zur Auswahl der Prozessgröße, welche am wahrscheinlichsten die Fehlerursache ist.

Abstract

Process control systems at production plants usually contain a large number of process variables. When detecting an abnormal process behavior, e. g. when passing a prior defined threshold or through multivariate fault detection, those process variables can generate an alarm. If the disturbance affects a major part of the plant this can lead to an alarm flood, as the process variables are interconnected to each other and a disturbance tends to propagate itself throughout the entire process. Operators need to select the most important alarms and need to make important decisions possibly under increasing stress, as well as time pressure in order to detect the root cause of the fault. To this end, it is of great benefit to automatically detect those process variables which are closest to the source of the disturbance and discriminate against those caused from a propagation of the original disturbance.

The proposed approach for the localization of plant-wide disturbances explored in this thesis is one that uses statistical relationships and time-shifts in the measurement data from the process variables, e. g. through dynamics in transfer functions or dead times, to track the disturbance propagation path and to detect the root cause of the fault. Methodologically, this means that cause-effect dependencies among process variables need to be detected. Process variables that show a strong causal impact on other variables in the process at a plant come into consideration as being the root cause.

For the automatic detection of causal dependencies in this thesis, several methods are proposed, which are based on cross-correlation, Granger causality, transfer entropy and support vector machines. Especially for support vector machines, a new approach has been developed. This new approach uses a variable selection that is based on time-shifted input signals to detect causal dependencies in measurement data. The detection of causal dependencies consists of a qualitative statistic and a quantitative part, that all developed methods have in common. This means that a significance test is initially performed to determine if a causal dependency exists among the investigated process variables.

Regarding the quantitative part in this work, the concept of *causal strength* has been developed to calculate the causal impact of one process variable on another. The causal strength is in that case a measure which quantifies the cause-effect dependency between two process variables. The main purpose is to describe the relationship for all methods between a value of 0 (no cause/effect) and 1 (sure cause/effect), whenever a significant causal dependency has been detected between two variables. The use of these quantitative methods results in a causal matrix, which describes the relationships for all used process variables and produces the resulting disturbance propagation path. Through a standardization between 0 and 1 it is possible to combine the resulting matrices of the different methods through averaging and to generate more robust results with respect to the localization of the root cause.

As each method has its specific assets and drawbacks, benchmark data sets were developed to perform an evaluation of the characteristics. These benchmark data sets cover aspects such as large dead times, nonlinearities, set-point changes, or data sets that consist of more than one root cause. To proof the concept, the proposed methods are tested on a laboratory plant as well as on an industrial glass forming process. The laboratory plant consists of an installation in which water is recirculated in two tanks. To test the methods, several experiments were performed. These experiments test the proposed methods against failures such as a loose electricity connection of the pump, an air pressure leak of the valve, and as an external fault the clogging of an inlet pipe. Through the averaging of the causal matrices it was possible to localize the root cause of the fault in all cases. Regarding the glass forming process, the methods were used to reconstruct the disturbance propagation path of a defect contained in the glass cylinder, which occurs during production.

In addition, several procedures have been developed for the visualization of the causal matrices to achieve a better analysis of the results. These visualizations express a comparison of the different methods on the same data set in the form of a bar chart, representing the disturbance propagation path as a partially directed graph and the generation of a root cause priority list to describe the process variable that has been identified as being the possible root cause.

Notations

A	Simplex
$\mathbf{Adj}_{X_i}$	Adjacent nodes of variable X_i
C	Cause
$C_{\text{thresh}}^{\text{CCF}}$	Significance threshold for CCF
D_{l}	Glass forming process lower diameter
D_{u}	Glass forming process upper diameter
D_i	Partitioning size of a random variable
$\boldsymbol{E}$	Set of edges in a graph
E	Effect
E_1, E_2	Activation energy for chemical reaction
$E_{\overline{U}_i Y}$	Prediction error restricted model
E_{test}	Residual sum of squares test data
E_{training}	Residual sum of squares training data
$E_{U_i Y}$	Prediction error unrestricted model
F	Fault, volume flow rate
$F(x)$	Strong classifier
Φ_{uy}	Set of variables used as input for SVM
$\Phi_{uy}^{\text{ranked}}$	Set of ranked variables from Φ_{uy}
$\mathbb{G}$	Graph
G^2	Test statistic for conditionally independency
$H(s)$	Transfer function
H_u	Entropy
$H_{u,y}$	Joint entropy
$H_{y|u}$	Conditional entropy
K	Length of time series

K_a	Amplification factor
K_{step}	Amplitude of step function
$L(\cdot)$	Lagrangian function
N	Number of performed runs, number of classifiers
N_r	Samples per rise time
P	Symptom pattern
$P(\cdot)$	Probability distribution
P_n	Process pattern
$Q^{\text{CCF}}, Q^{\text{TE}}, Q^{\text{GC}}, Q^{\text{SVM}}$	Causal matrix for proposed method
Q_{fus}	Combined causal matrix
$Q_{\text{fus}}^{\text{pk}}$	Combined causal matrix with a priori knowledge
RC	Weight of being the root cause
$\boldsymbol{S}$	Subset of $\mathbf{Adj}_{X_i}$
T	Time constant
T_{d}	Signal dead time
T_{sn}	Nyquist frequency
T_{s}	Sample period
T_{sn}	Noise sample time
TE_{uy}	Maximum amplitude of $\text{TE}_{uy}^{\star}[\lambda]$
$\text{TE}_{uy}^{\text{thresh}}$	Threshold for causal significance
$\text{TE}_{uy}^{\star}$	Amplitude of the transfer entropy
$\boldsymbol{V}$	Set of nodes in a graph
R	Universal gas constant
V	Volume of cstr
$V_{0.632}$	Loss function from 0.632 bootstrap algorithm
X	Feature space
X_i	Random variable, process variable
$\alpha_i, \hat{\alpha}_i$	Support vectors
α_n	Weight of classifier
α	Threshold for statistical significance
$\beta_{\text{CCF}}, \beta_{\text{TE}}, \beta_{\text{GC}}, \beta_{\text{SVM}}$	Causal matrix fitting parameter
c	Compression rate, tuning parameter for notch filter
$c_{\text{in}}, c_{\text{A}}, c_{\text{B}}, c_{\text{C}}$	Measured chemical concentrations in cstr
$\hat{c}_{uy}$	Amplitude of CCF

d	Durbin-Watson test statistic, Tuning parameter for notch filter
ε, C, σ	SVM fitting parameters
$\varepsilon_{\mathrm{opt}}, C_{\mathrm{opt}}, \sigma_{\mathrm{opt}}$	Optimized SVM fitting parameters
ϵ	Residuals for Durbin-Watson statistic
ϕ	Size of subset for selected variables from $\Phi_{uy}^{\mathrm{ranked}}$
f	Frequency
$h_n(x)$	Weak classifier
$k(\cdot,\cdot)$	Kernel function
k_1, k_2, k_3	Chemical reaction speed
λ	Time delay between two time series
λ^{max}	Indice for CCF algorithm
μ	Linear mean
n	Model order, lag vector, selected classifier
ω_n	Notch frequency
ω_o	Natural frequency
p	Probability
$q_{X_i \to X_j}$	Causal strength
$\hat{\rho}_{uy_\lambda}$	Time shifted correlation coefficient
$\hat{\rho}_{uy}$	Pearson correlation coefficient
r	Set of process variables, residuals
σ	Standard deviation
σ^2	Variance
ϑ_{b}	Glass temperature
ϑ_{c}	Cylinder temperature
ϑ_{fl}	Temperature of fluid flow
t_{score}	Test statistic for t-test
u	Input time series
u_λ	Shifted input time series
u_π	Permutated input time series
v_{s}	Pulling speed
y	Output time series
ζ	Damper constant
AIC	Akaike information criterion

ARMA	Autoregressive moving average
BIC	Bayesian information criterion
CCF	Cross-correlation function
CRISP-DM	Cross industry standard process for data mining
GC	Granger causality
KDD	Knowledge discovery in databases
LOO	Leave-one-out error
MIMO	Multiple input multiple output
MPC	Model predictive control
LTI	Linear-time-invariant system
PC	Peter and Clark algorithm
PCA	Principal component analysis
PDAG	Partially directed acyclic graph
PICA	Piecewise constant approximation
ROC	Receiver operating characteristic
RSS	Residual sum of squares
SVM	Support vector machines
TE	Transfer entropy
VAR	Vector autoregression
cstr	Continuous stirred tank reactor
do	do-operator
df	Degree of freedom
pdf	Probability density function
psd	Power spectral density
sgn	Sign function

Contents

1

Introduction

This chapter gives a short introduction into the subject of data-driven fault localization and states the main aspects which motivate this work. The chapter closes by outlining the main contributions as well as explaining the organization of the thesis.

1.1 Motivation

Modern industrial plants are complex systems that need to run over several weeks or in some cases even months. During a production run, operating conditions can change which then can lead to unwanted abnormal behavior of the process. Hence, counteractions need to be undertaken to attain an efficient production. Changes in process conditions, such as wearing or clogging of the equipment or a change of some external condition like outside temperature can cause disturbances in the process. These disturbances have the potential to lead to faults. To resume production a fast elimination and restoration of the original process conditions is one of the highest priorities. Because modern plants' control and measurement devices are cross-linked with each other, a failure in major equipment can potentially lead to plant-wide disturbances resulting in a flood of alarms, which makes it difficult to find the root disturbance. An illustrative example is given in figure 1.1 which shows the scheme of a chemical plant where a fault in a valve, marked with a red flash, becomes a plant-wide disturbance and generates the previously mentioned flood of alarms. As not all relations of the different process parameters are well known, the detection of the source of a fault and the reconstruction of the corresponding disturbance propagation path is not an easy task. Often, a correct fault diagnosis mainly depends on the experience and the expert knowledge of the process operator. Besides being complex, modern processes have the advantage that they are highly automated and are largely equipped with measurement devices. Therefore, by using process data containing the disturbed measurements, it is possible to localize or at least to narrow down on the cause of a disturbance.

Because of its importance and benefit, performing a data-driven detection of faults in process data is the main topic of the thesis. As it takes some time for the disturbance to propagate through the process, the idea is to use temporal information as well as the calculation of statistical relations among the different process variables to reconstruct the disturbance propagation path. The node representing the root of the calculated path then represents the process variable which is the source of the fault or the process variable that detected the fault first. There are two points of view on how to motivate the development of these methods, namely in terms of process supervision and in terms of data mining. This is explained below.

View-point of process supervision   Following the study of the NAMUR consortium presented by Hagemeyer [HP09] in 2009, 78 % of the process engineers are not satisfied with the alarm management of their processes and wish to have further research in this direction. The concept of state-of-the-art alarm management for process supervision is to give an alert to the operator if one or several process devices deviate from their normal conditions. As a process consists of a large number of different process devices that provide an alert when the process shows an abnormal operating situation most of the alarms are not meaningful and can be ignored by the operators. Despite this, when an alarm flood is generated through a fault in a piece of major equipment, operators need to analyze the situation quickly and need to make important decisions to prevent damage to the plant. Therefore, a more intelligent alarm management is necessary that selects the most meaningful alarms and is capable of bringing an operator's attention to the source of the fault. Regarding the chemical plant

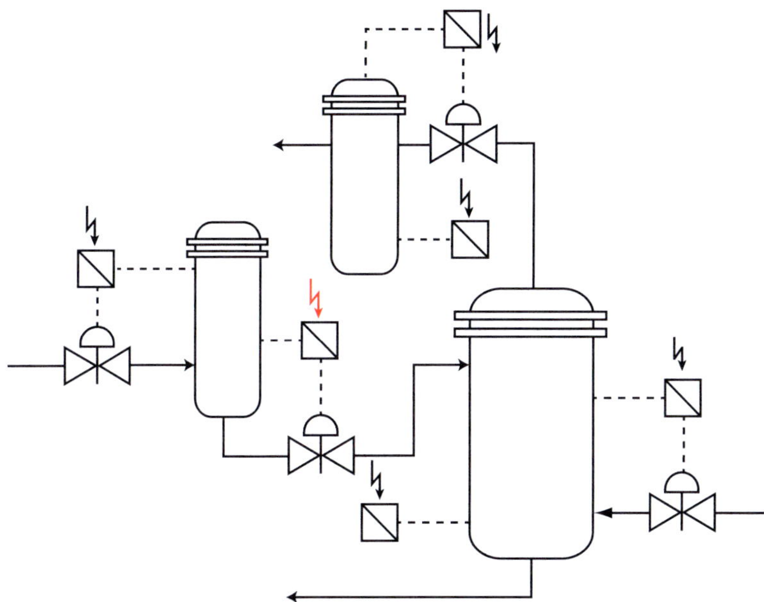

Figure (1.1) A fault in one major process equipment, marked in red, generates disturbances that propagate through the process generating various alarms. Localizing the
root cause and the disturbance propagation path helps resolving the problem.

in figure 1.1 insufficient alarm management means that six alarms alert while
only the alarm of one valve is meaningful. Therefore, data-driven methods
for the detection of cause-effect dependencies can help to improve the alarm
management by giving the five meaningless alarms a low priority.

View-point of data mining   Having almost unlimited storage space and a
high level of computational power, engineers have begun to think about how to
analyze process data further. One of the first concepts was defined by Fayyad
[UGP96] in 1996 named Knowledge Discovery in Databases (KDD), wherein
an iterative procedure is suggested on how to extract new knowledge out of
available data sets. Later, in the year 2000 several industrial companies formed
the Cross Industry Standard Process for Data Mining (CRISP-DM) [She00b]
which describes not only the analysis, but the complete life cycle of a data

mining project. In both cases, the essential part is to use statistical methods to find patterns in data. Regarding industrial processes these methods are mainly used for fault detection tasks but not for diagnosis. The main reason is that the learned patterns usually correspond to black-box models, meaning that the internal operation of the model corresponds to some mathematical function but does not necessarily have a physical relation to the process. Developing methods that detect cause-effect relationships in process data can give further insights into the physics and the interconnection of the plant. Therefore, the main motivation from the point of view of a data miner is to develop methods that do not only represent some statistical relations but describe causal dependencies to generate new knowledge by analyzing the process data.

1.2 Related work

This section gives a brief review to the field of data-driven fault analysis which is most relevant for this thesis. Localizing faults in a process can be seen as a part of fault diagnosis performed in process supervision. A more detailed introduction to process supervision with a thorough literature review will be given in chapter 2.

In general, approaches for root cause localization can be separated into two parts, namely analytic and data-driven methods. The difference of both approaches is that analytic methods consist of a mathematical process model and compare model parameters against measurements, while data-driven methods are based on recent process data. As introduced in section 1.1 only a small amount of real plants is modeled analytically, which gives the motivation for this thesis to explore data-driven methods for the localization of faults. These methods are based on the calculation of statistical relationships and time-shifts in the measurement data from process variables. Methodologically, this means that cause-effect dependencies among process variables are calculated and variables that show a strong causal impact on other variables come into consideration as being the root cause of the disturbance.

Several methods have been already developed to test for causal dependencies in data. One of the first approaches was made by Granger [Gra69] who compares two autoregressive models to each other. In so doing, the first model contains only lagged values of itself and the second model is augmented with lagged values of another variable. If the augmentation helps to increase the regression, it is assumed that this variable is the cause of the other. As another approach, Horch [Hor00] proposes an algorithm based on the cross-correlation function for causal analysis. In that case, the absolute maximum value of the cross-correlation function and the corresponding time shift are used for the description of causal dependencies in the data. Schreiber [Sch00] presents a concept to interpret causal dependencies as the information flow from one variable to another. Therefore, he defines the concept of transfer entropy, which measures the reduction of uncertainty when assuming that one variable predicts future values of the other. There are further methods existing to detect causal dependencies. Bayesian networks [EM07] have been extended to detect dynamic behavior in data, which again can be used to calculate causal dependencies. Bauer [Bau05] proposes several algorithms which use a nearest neighbor approach as a causal measure by predicting the value of one variable by another.

This shows that already various methods exist to detect causal dependencies in a data set while each of the introduced methods has its own characteristics and one method can complement another. Hence, it can be concluded that the performance of the methods depends on the occurred disturbances and the characteristics of the underlying process. Therefore, Yang [YX12] proposes always to try different methods on the same data set to obtain reasonable results.

The localization of faults and disturbances builds the application of the data-driven methods in this thesis. A fault is commonly understood as a malfunctioning of one or several process devices. This fault leads to disturbances in the process which affects other process devices and reduces the overall performance of the plant. There exists an almost unlimited variety of possible faults and disturbances occurring in a process. Since there exists this large variety, recent research focuses on the development and analysis of disturbances in industrial case studies. In that area recent research has been made by Bauer [Bau05] who describes common faults in chemical plants, Singhal [SS05] who describes disturbances occurring from valve stiction and Horch [Hor00] who focuses on disturbances occurring from dead-zones and friction.

In a plant with a large number of process variables, the disturbance that occurs at a certain position in the process, eventually becomes a plant-wide disturbance. In this work, the localization of the origin of the plant wide disturbances and the reconstruction of the disturbance propagation path is a central part for the analysis of the data-driven methods. If the path can be reconstructed, the source of the disturbance can be found, which again helps to identify the cause of the occurred fault. Hence, the relevance of the disturbance in this thesis is to test if the data-driven methods will be able to track back how it propagates through the process.

1.3 Contributions

The introduction showed that various methods already exist to detect causal dependencies in a data set while each of these methods has its own characteristics. In the course of this, important aims of this thesis are a concept that makes these methods comparable to each other and the development of a design approach which combines them to one causal matrix. Another important goal is the proposal of a new method for the detection of cause-effect relationships based on support vector machines. Furthermore, benchmark data sets are developed, which identify the advantages and drawbacks of the methods. These benchmarks cover artificial data sets as well as real process data. In detail, the main contributions of this thesis are:

- A method which uses support vector machines for the calculation of causal dependencies. The novelty of this method is a recursive variable selection and model reduction approach (section 3.7);

- Improvement of a method based on the cross-correlation function by using permutation tests for the detection of significant causal dependencies (section 3.4). Transfer entropy and Granger causality are extended by calculating a quantitative value which makes all proposed methods comparable (section 3.5 and section 3.6);

- Development of a design approach which combines all proposed methods to one resulting causal matrix. This is presented without a priori knowledge and when including a priori knowledge in terms of known causal dependencies (chapter 4);

- Development of an artificial benchmark data set for a detailed evaluation of the characteristics of the proposed data-driven methods (section 3.2);

- Application of the developed design approach on an experimental laboratory plant while introducing different faults (chapter 5) and introduction of a new industrial case study. The functionality of the methods is verified on an industrial glass forming process when analyzing the disturbance propagation path of a weld (chapter 6);

- Study on the setup of the data acquisition system. Guidelines are established for the selection of the sampling period, data compression and the filtering of oscillatory data when calculating causal dependencies (chapter 7);

- Development of a visualization method for the causal matrices. This visualization includes a comparison of the methods, the combined causal matrix and the detected disturbance propagation path (section 3.3.1);

Publications The work presented in this thesis resulted in several publications. A summary is given in chronological order. How to use data-driven methods for process supervision is explained in [KMB09]. The methodology how variable selection and model reduction can be performed using support vector machines is presented in [KB09a] and [KB09b]. Using support vector machines with a focus on process supervision is presented in [KBF10] and [KB10a]. An overview how to use probabilistic causal methods for the detection of the causal structures in production data is given in [KB10b]. Using these methods with a focus on the detection of the reaction scheme of the chemical stirred tank reactor is given in [KBF11]. A survey of the suggested methods in this thesis is given in [KGHM11] and with a larger focus on visualization in [KGHM12].

1.4 Thesis Organization

The thesis includes eight chapters which are all based onto each other. **Chapter 2** starts with initially reviewing the main aspects of process supervision and describes potential disturbances that can occur in a plant. Next the fundamental difference of correlation vs. causality and two different perspectives of causality are explained. Finally, it is illustrated how probabilistic causal methods can be used to analyze production data. **Chapter 3** begins with classifying causal discovery methods using temporal information depending on their algorithmic framework. Next, several benchmarks and visualization methods for the representation of causal dependencies are proposed. The core of the chapter covers the explanation of the proposed methods, namely cross-correlation, transfer entropy, Granger causality and support vector machines. All methods are thoroughly tested on benchmark data sets. **Chapter 4** summarizes the results from the benchmark data sets by suggesting a sunburst graph for the selection of the correct method depending on the underlying process. Next it is proposed to combine all methods into one resulting causal matrix by assuming that all methods work equally well on a data set. Finally it is shown how the disturbance propagation path of a continuous stirred tank reactor can be calculated with and without including a priori knowledge. **Chapter 5** shows the results of the suggested methods when being tested on a laboratory plant in which water is pumped around in cycles. In total six different faults are investigated, namely a loose electricity connection in the pump, a valve air pressure leak, faults in pump and valve, tube clogging, a loose electricity connection in an oscillating pump and a faulty level sensor. **Chapter 6** investigates the methods on a known disturbance of an industrial glass forming process. The process characteristics are explained and the disturbance propagation path is calculated using productions containing this disturbance. **Chapter 7** investigates the impact of the sampling period, data compression and oscillations when having measurement data available. Furthermore, the impact of filtering is investigated when removing oscillations in signals by using notch filters. **Chapter 8** summarizes the results containing a critical discussion and describes open issues for further research.

2

Disturbances in Process Data

This chapter gives a thorough introduction how causes of faults can be localized when having disturbed process data at hand. Therefore, in this chapter the state-of-the-art of process supervision is reviewed and different types of faults and disturbances that can possibly occur in a plant are explained. Thereafter, the fundamental differences of causality and correlation are explained and the different possibilities on how to detect causal dependencies in acquired measurement data are demonstrated. The chapter closes by illustrating how the concept of probabilistic causality can be used to detect the cause of a varying product quality in a continuous stirred tank reactor.

2.1 Process Supervision

As processes become more and more complex monitoring the performance of a process is of crucial importance to attain an efficient production. Conditions during a production can change which can lead to unwanted behavior of the process and counteractions need to be undertaken. Hence supervising (or monitoring) a process needs to fulfill several tasks. In first place this means that the present state of the process is indicated to the operator and that undesired or unexpected behavior is reported to him. Still supervision is related to more than only performing a condition-monitoring of the process behavior as supervision addresses all aspects needed to take appropriate actions to prevent the process from damage or failure. There are several books about process supervision available while one of the first ones was written by Himmelblau [Him78] who focusses on chemical plants and Pau [Pau81] who gives a more general introduction. More recent literature was written by Chiang [CRB01] which focusses mainly on statistical process analysis and Gertler [Ger98] which gives a more general view about the topic. In the present work the review given for process supervision mainly focusses on the framework developed by Isermann [Ise06]. He separates supervision into three different levels, namely monitoring, automatic protection and supervision with fault diagnosis. All three levels are sketched out in figure 2.1 and are explained below.

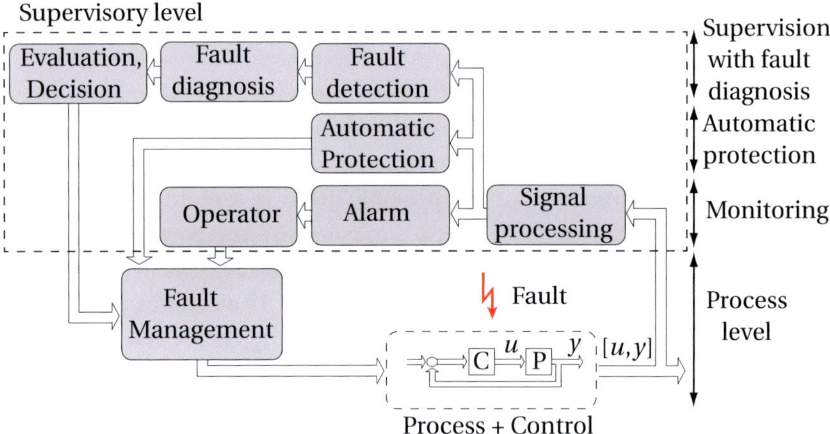

Figure (2.1) Illustration of the different levels of process supervision as explained in the framework of Isermann [Ise06].

- **Monitoring:** Monitoring the condition of a process means that it is checked if the values of the process variables stay inside some predefined range of tolerance. If a measurement leaves its tolerance, e. g. by passing a threshold, an alarm is generated and displayed to the operator. In a next step the operator has to decide which action needs to be undertaken to avoid a possible damage to the process.

- **Automatic protection:** Depending on the occurred fault the supervisory system automatically performs some counteraction and does not wait for an input coming from the operator. In most cases an automatic counteraction means that the process is automatically shut down or set into a fail-safe mode. The advantage when setting the process into a fail-safe mode is that production is still sustained by taking into account the cost of low efficiency.

- **Supervision with fault diagnosis:** Performing a process supervision with fault diagnosis covers three steps. In the first place, the fault needs to be detected and symptoms that detail the influence of the fault in the process need to be specified. Next, a fault diagnosis is undertaken that describes the kind, size and location of the fault [Ise06]. Finally, the diagnostic needs to be evaluated and appropriate decisions have to be done either by the operator or automatically. The concept of fault diagnosis is reviewed in further detail in section 2.1.2.

All levels of supervision can be conducted locally, meaning that faults are detected by monitoring process variables on a one-by-one basis or more globally by combining several variables into one value which is then used for supervising. Independent of the selection of the supervision level, if a fault is detected, some counteraction needs to be conducted. As illustrated in figure 2.1 this is covered under the synonym fault management. Faults in the process can potentially lead to a failure or damage. Hence a fast detection and the selection of the correct counteraction is unconditionally necessary. Therefore, Isermann [Ise06] defines several actions that need to be accomplished depending on the hazard class of the fault. In detail these are safe operation (e. g. a shutdown), reliable operation (e. g. switching to a safe production mode), reconfiguration (e. g. switching to another sensor system), inspection (e. g. detailed diagnosis of the fault), maintenance (e. g. exchange of fault producing process components) and repair (e. g. revision of process components). Process supervision is performed in a wide area of industries. Ragot [RM06] proposes methods for process supervision for urban water supply, Kano [Man08] for applications in steel industry, Geng [GZ05] in chemical engineering and Pinder [PG93] for food industry.

2.1.1 Fault Detection

Prior to performing a diagnosis of a fault, the abnormal process behavior needs to be detected. According to Isermann fault detection can be separated into three different categories, namely knowledge-based, signal-based and process-model-based.

- **Knowledge-based:** A knowledge-based fault detection means that analytic and heuristic symptoms are generated and used to detect abnormal behavior in the process. Analytic process information addresses all quantifiable symptoms resulting from signal analysis by using characteristic values such as frequencies, amplitudes or variances. Heuristic symptoms cover all information that can be generated through operator knowledge. Information coming from operators includes geometry, noise, special colors of a product or concerning process history for example the last revision, wear or the knowledge of former faults. Both heuristic and analytic symptoms can be summarized to generate one unified representation of symptoms and this representation can be transformed into linguistic terms to monitor the process and to detect faults.

- **Signal-based:** Detecting faults based on analyzing a single signal can be done in its basic form by performing limit checking, trend checking or change detection. Limit checking means that a threshold is defined and if the measured variable crosses this value a fault is detected. Trend checking works in the same form with the difference that not the signal, but the derivative is calculated to detect a fault. The concept of change detection is to calculate the mean and variance of a signal over a certain time frame. If mean and/or variance differ significantly from an expected value a fault is detected in the process. Furthermore, recently machine learning methods and multivariate statistics found their ways into plant-wide monitoring, which is of special importance when supervising large-scale processes. In detail the concept is to project the original process data onto a lower dimension and thus to reduce the number of variables to monitor. Of particular importance is in that case the principal component analysis (PCA), which makes an axis transformation of the original data to keep as much variation as possible in the data set when the dimension is reduced. In terms of machine learning a promising approach is established by using self-organizing maps which are a special type of neural networks (see e. g. [Fre08]).

- **Process-model-based:** The concept of a process-model-based fault detection means to use a mathematical process model for monitoring. This model comprises the main parameters of the process and describes its essential physical relations. Having the measured input and output signals at hand the process-model-based fault detection can be made by calculating either parameter estimates or residuals. The idea of parameter estimates for fault detection is to monitor the deviation of one or several estimated process parameters while the process is running. A significant difference from the original estimates indicates a fault in the process. The approach based on parameter estimates has the advantage that the determination of the variation of process parameters already gives an insight of the type of the fault and makes a localization and diagnosis easier (e. g. when monitoring the stiffness of a spring). Another concept, if the process parameters are known, is based on developed state observers which describe the process. In this approach the residuals of process and those from the observer are compared and if the residuals increase while the process is running, a fault is indicated.

Figure 2.2 gives a survey of some of the most used methods applied for fault-detection in processes, while distinguishing them into data-driven and analytic approaches. There are several other ways on how to organize the methods. Chiang [CRB01] classifies the methods concerning data-driven, analytical and knowledge-based, Isermann classifies the methods besides the classification used above into single signal and multiple signal methods. A priori knowledge can be incorporated into all methods, e. g. through the definition of thresholds when a fault is detected.

Process modeling in practice   As the present work is devoted to data-driven methods for fault localization, which do not need a process model of the plant,

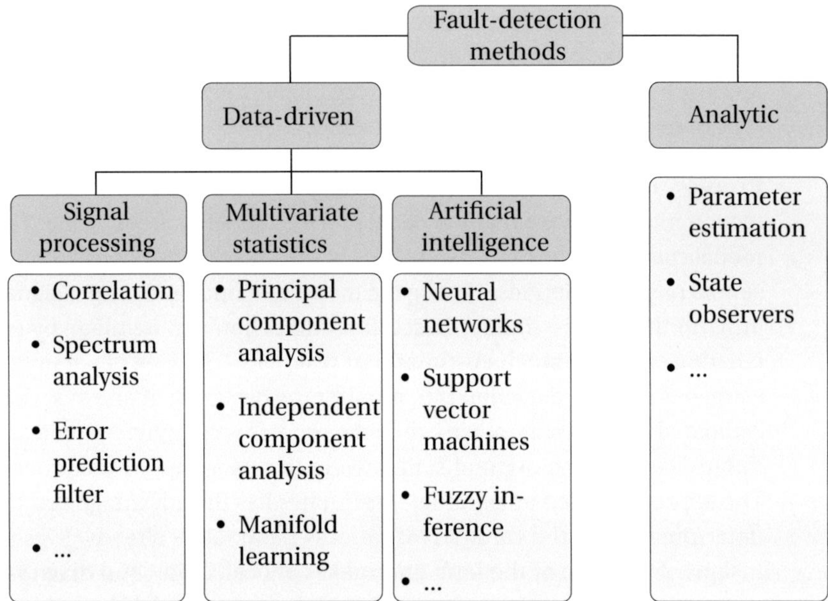

Figure (2.2) Survey of possible fault-detection methods classified in data-driven and analytic approaches. In this thesis the focus is on data-driven approaches.

it is important to know how many processes are modeled in practice to check if the methods in this thesis are needed in industry. In this case, studies investigating how often model predictive control (MPC) is performed in practice can be used as a source of information. An investigative study of about 10.000 plants made by Dittmar [RP06] in 2006 revealed that in most industrial sectors less than 10 percent of the processes use MPC. Thereby refineries with 49.2 % are in the majority, in chemical industry it is 15.6 % and in the other industrial sectors e. g. paper industry, power plants or coal mining less than 5 % use MPC. From this it can be concluded that still the majority of industrial processes is not physically modeled meaning that process-model-based observers for process supervision cannot be used without a large amount of work. Another market study made in 2009 by the NAMUR consortium [HP09] analyzing the expectations of future development in process engineering gave similar results.

Having a continuous process, the wish to use MPC in the future was around 25 %, having a batch process the wish to use MPC was 15 %. To summarize, all surveys show that still the majority of the plants is not modeled and that there is a need to use data-driven methods for fault detection and localization.

2.1.2 Fault Diagnosis

As outlined in figure 2.1 fault diagnosis is done after a fault has been detected in the process. Following the definition of Isermann [Ise06] a fault diagnosis means the

'Determination of kind, size, location and time of detection of a fault
by evaluating symptoms.'

As in this work the main focus is on fault localization and the calculation of the disturbance propagation path the suggested methods in chapter 3 can be seen as part of the diagnosis of a fault. Performing the complete diagnosis of a fault can be done either through classification or inference methods. Both are illustrated in figure 2.3 and summarized for completeness.

- **Classification:** The concept when using classification methods for fault diagnosis is to use an experimentally generated knowledge base. In that case a process pattern P_n is determined when having normal process behavior and several faults are applied experimentally to generate a set of symptom patterns P, resulting from the fault detection method. When a possibly new fault F is detected, by comparing P and P_n a fault diagnosis can be concluded by using the generated knowledge base.

- **Inference:** If there are some basic relations known, this a priori knowledge can be used for the generation of rules and then used for diagnosis. These rules can be described through simple if-else relationships in the form of `if <cause> then <effect>` or constructed in terms of a decision tree. As another approach it is possible to construct Bayesian Networks and to interpret the edges as being causal (see[Mur02]). The advantage by following this approach is that one deals with a probabilistic graphical model and therefore, probability theory can be applied for fault diagnosis.

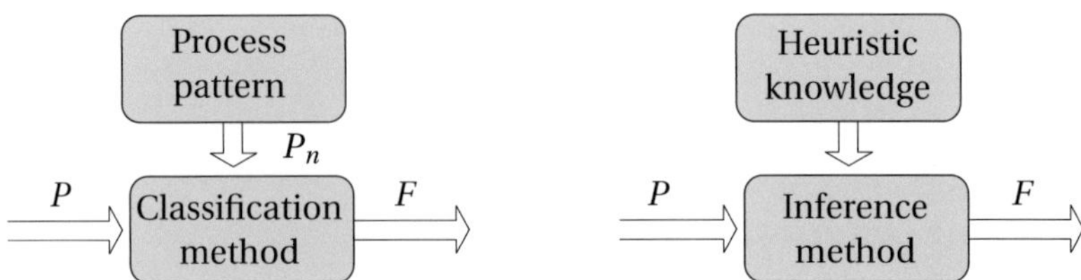

Figure (2.3) Methods used for fault diagnosis based on pattern recognition (classification methods) and process knowledge (inference methods).

2.2 Disturbances and Faults in Process Data

Following the definition of Isermann [Ise06] a fault is defined as an

> '*Unpermitted deviation* of at least one characteristic property of the system.'

In other words, this means that at least one process component differs significantly from its expected behavior. By definition process components cover the whole range of devices (e. g. valve, pumps, pipes) that are used in process technology. Additionally, Isermann [Ise06] distinguishes faults from disturbances and therefore defines a disturbance as

> 'An *unknown* (and uncontrolled) *input* acting on a system.'

Based on this definition, it can be concluded that when having a fault in the system, disturbances will occur due to the deviation of the faulty process component from its normal behavior. These disturbances can then be measured by acquiring the process variables. To put it in another way, disturbances are the symptoms of faults in a process and a fault can be detected and located by recognizing abnormal system behavior generated through the disturbances. As mentioned, in the present work the main focus is on the localization of the source of the fault. Therefore, the main idea is to perform a backpropagation of the disturbances that have an effect on the different process variables and finally to find the underlying cause of the event that generated the measured disturbances. The cause of this event is commonly called the root cause and the way the disturbance travels from one process variable to the next one is called the disturbance (or fault) propagation path. Finally, by finding the root cause, this means that the faulty process component that generated the disturbances is localized.

2.2.1 Disturbance Propagation Path

The disturbance propagation path describes the way a disturbance travels through the process. In that case the knowledge of the propagation path can significantly improve the fault localization as the path gives information about how the fault affects the behavior of the process. For the reconstruction of disturbance propagation paths two sources of information, namely temporal information and attenuation can be used. Additionally, in some cases it is possible to reconstruct the disturbance propagation path using probabilistic calculations. In the temporal case time delays are used as causal information for reconstruction. An example is given by a water tank which can be described as a 1st-order system or as another example a measurable dead time between two process devices. How to use temporal information in connection with the statistical methods for the calculation of a disturbance propagation path will be explained in section 2.3.2. Another possibility to calculate the disturbance propagation path is to analyze production data using probabilistic methods. The leading thought in this approach is to calculate conditional dependencies among process variables and to interpret the found dependencies as being cause-effect relationships. Interpreting the dependencies as being causal can be used when analyzing a set of production data for example when analyzing the impact of high temperature combined with a weak fluid concentration on the resulting product quality. This approach is explained and investigated in further detail in section 2.3.1.

Another way to calculate the disturbance propagation path can be done by comparing the magnitude of the disturbance in each measurement device. The main idea is that the magnitude of a disturbance usually decreases further away from the root cause while the frequency spectrum is not affected. Using attenuation for the calculation of the propagation path is explained in further detail in [Bau05] and will not be investigated in this work.

When calculating a disturbance propagation path, it is important to notice that a disturbance does not necessarily travel into the direction of the gas or fluid flow. A basic example is given in terms of a faulty valve that closes unexpectedly while having a sensor that measures the pressure of the fluid before passing the valve. As the valve closes, the pressure in front of the valve increases as the fluid cannot flow through the process meaning that the disturbance propagation points into reverse direction.

2.2.2 Possible Faults in a Process

There is an almost unlimited variety of possible faults that lead to disturbances in measurement data. Common faults in chemical plants have been investigated by Bauer [Bau05], who bases her analysis on an earlier one made by Desborough and Hill [DM98]. She divided the types of causes into three different groups, namely process problems, controller tuning problems and valve problems. Process problems cover the impact of the process configuration, controller tuning problems address inappropriately tuned control modules and valve problems deal with the problem of wearing in valves. There are further studies available that deal with the problem of how faults occur in systems. Singhal [SS05] analysis the causes of oscillations coming from valve stiction and describes a possibility to detect it. Horch [Hor00] analysis the impact of static friction, dead-zones and further nonlinearities as possible root causes of faults in a process. In the present work a distinction of different faults is made concerning physical constraint problems, external disturbances acting on the system, actuator problems and sensor problems. Each type of fault is determined either in simulation, on the laboratory plant explained in chapter 5 or on the industrial glass forming process in chapter 6. In detail these types of faults can be described as follows:

- **Physical constraint problems** These are caused due to the nonlinear behavior of physical reaction schemes. Chemical or biological reactions are often nonlinear or discontinous. An example is that in some cases only a small change in some fluid concentration can have a large impact on the quality of the resulting product. Another example is the large impact of temperature. In some reactions a small temperature change can lead to large variations in the final product. As these problems can be caused by various process devices, knowing the root cause device and the underlying disturbance propagation path is extremely useful. The analysis of this kind of problems is investigated on a simulated continuous stirred tank reactor.

- **External problems** This type of faults is not caused through a malfunctioning of a process device. Typical examples are leakage or deposit in pipes that lead to a measurable flow or pressure reduction. Another example of external problems are faults that are already present in a preproduct of the process. In this case it is reasonable to perform a fault localization to detect the first process device that is influenced by the disturbance. As in some cases the fault is already known, it is extremely helpful to be aware of the disturbance propagation path to understand the impact of the fault on the plant. This type of problems is analyzed on the laboratory plant and on the industrial glass forming process.

- **Actuator problems** This type of problems addresses disturbances coming from the malfunctioning of an actuator. There is a wide variety of possible actuator faults occurring while a system is running. Examples are electricity drop-outs in pumps or other electric drives. In terms of valves examples are stick-slip, clogging, leakage or evaporation. Problems like these are excessively investigated on the laboratory plant.

- **Sensor problems** These faults are caused due to a malfunctioning of some measurement device. Faults like this can only become plant-wide if the measurement value of the sensor has an impact on the behavior of an actuator. In this thesis sensor problems are investigated on the laboratory plant.

2.2.3 Analyzed Disturbances

Disturbances in process data can have different forms. Obviously not all disturbances can be detected as easily as others. Therefore, four common disturbances are selected for investigation in the benchmark data which are all outlined in figure 2.4. In detail this is white noise, colored noise generated using a 1st-order system, a step function and a sinusoidal oscillation. The impacts of white noise, colored noise and the step function are evaluated on the benchmark data sets presented in section 3.2. As the sinusoid is a cyclic signal that doesn't carry causal information it is investigated how filtering the signal affects the resulting cause-effect dependencies. The impacts of oscillations and filtering them when calculating the disturbance propagation path are determined in chapter 7.

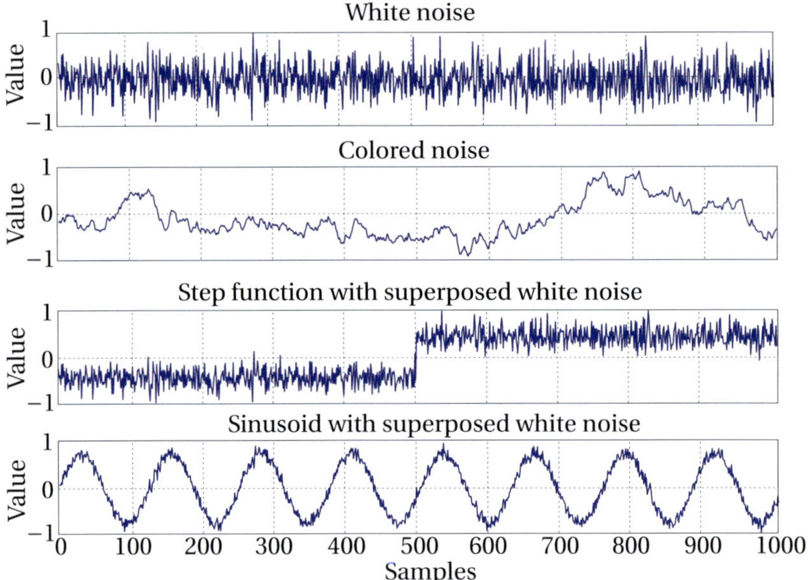

Figure (2.4) Characteristic disturbances occurring in a process that are investigated in the present work. Different data-driven methods are tested for the reconstruction of the disturbance propagation path using these signals as root cause.

- **White noise** A time series that consists of white noise means to have a sequence of uncorrelated random variables having constant mean and variance. In system identification this signal is often used as it has equal power over a fixed bandwidth. White noise is used in the benchmark data sets to investigate the behavior of the methods when having different system configurations (e. g. nonlinearities or higher-order systems). Having white noise is the idealized form of a disturbance and is especially used to compare the behavior of the different proposed data-driven methods on the system configurations. On real plants disturbances are always autocorrelated up to a certain point.

- **Colored noise** Compared to white noise colored noise has a limited bandwidth over its frequency spectrum. Having a colored disturbance means that the disturbance contains information about itself in past values and future values can be predicted from it up to a certain point. This can lead to difficulties when calculating the disturbance propagation path as the suggested methods need to subtract the information the signal contains about itself to detect causal dependencies.

- **Step function with superposed white noise** Having a step function means that the mean value of the signal changes abruptly and the methods have to calculate the disturbance propagation path while dealing with a non-stationary disturbance. As all methods assume stationarity, the robustness of the methods when violating this assumption is investigated. The impact of a step function is tested on the benchmark data sets and on the laboratory plant. In detail, on the laboratory plant the step function is used to generate an external disturbance and a joined error in pump and valve.

- **Sinusoid with superposed white noise** A sinusoid is the prototype of a periodic disturbance. As it represents a completely cyclic oscillation no causal information can be drawn from it and all the causal information is hidden in the superposed noise. In other words, having a sinusoidal oscillation the frequency of the sinusoid can be suppressed for reconstructing the fault propagation path. This is analyzed in further detail in chapter 7.

2.3 Measuring Causal Dependencies in Data

The core of the reconstruction of disturbance propagation paths is to detect cause-effect relationships in a data set. In general terms causality is described as the relationship between a cause and an effect, while the effect is always the consequence of the cause. To give a short historical review, the debate about causal reasoning starts in Western philosophical tradition with Aristotle [Lea88] and rests until today. Aristotle argued that to understand an object completely, one has to understand its 'why' meaning to understand the cause why the object exists. Later on in the eighteenth century the philosopher David Hume [May70] came up with a new definition of causality where he defined that it is not possible to distinguish between cause and effect except one develops a habit in mind that associates two events with each other that are always occurring one after another [Hum96]. Therefore, Hume describes a list of eight numbers with which it is possible to decide which one of two events is the cause of the other. One of the main points is that cause and effect need to be interconnected in space and time, the cause must be prior to the effect and that the same cause always leads to the same effect. In modern times several fields reaching from economics, psychology, history, religion or law deal with the notion of causality while all fields have a different view on the definition of causality.

View-point of engineering   Seeing causality from an engineering point of view is very precise. Engineers want to change something on a system which will result in a desired outcome. Applying a causal analysis in terms of fault localization means that engineers want to find the root cause of a disturbance, while applying a causal analysis regarding process optimization means that engineers want to find the optimal operating points of a system, which e. g. will result in a better product quality. In both cases a data analysis based on the detection of causal dependencies in the process is indispensable as an analysis finding only correlations in the data does not describe cause-effect relationships. In other words, for the engineer without having a thorough expert knowledge and being only aware of correlations among the process variables, it is not possible to tell how the results of a correlation analysis will have an effect when modifying the process (e. g. when changing the process control). The following section is devoted to the explanation on how to gain causal knowledge from an acquired data set, separated into probabilistic causality and causality with time dependency.

- **Probabilistic causality** When the acquired data set consists of measured values over several productions each production can be described as the result of an independent experiment. In that case methods can be applied that find cause-effect relationships based on the calculation of conditional dependencies in the data. There are essentially two different approaches which can possibly be used, namely one that uses statistical tests [Pea00] and one that performs a search-and-score approach [Mur02]. Since the search-and-score approach only uses some quality function to weight the found relationships between the variables, it does not ensure that those are in terms of the causal direction correct. Therefore, the approach using statistical tests will be explained and investigated in further detail in the following section.

- **Causality and time** If the acquired data consists of different time series, temporal information in terms of system dynamics and dead times can be exploited for analysis. These approaches are all based on the calculation of statistical dependencies among the measured time series while performing time shifts in the data. In total there are four methods, namely the cross-correlation function, transfer entropy, Granger causality and support vector machines investigated in this work.

2.3.1 Probabilistic Causality

Compared to the philosophical definition, which says that the same cause always leads to the same effect, probability theory can be used to weaken this definition. In detail the central idea is that a cause C defined as $C : \Omega \rightarrow \{0,1\}$ increases the probability that an effect E defined as $E : \Omega \rightarrow \{0,1\}$ occurs. In that case the value 1 means the event has happened and 0 that it did not occur.

An illustrative example in process engineering is that high temperature in a plant increases the probability to have a product with low quality, but not all low quality products are the outcome of high temperature. Writing this in a probabilistic way means, that C can only be a cause of E, if $P(E = 1|C = 1) > P(E = 1|C = 0)$. Still Pearl [Pea00] points out that an increase of probability cannot be described purely from observational data, but through an external intervention. This means that if one wants to be sure if C has an effect E one has to conduct experiments on the process. Therefore, Pearl introduces the do-operator, that forces to fix the cause C on 1. In probabilistic notation this means that C has a causal influence on E, if $P(E = 1|\operatorname{do}(C = 1)) > P(E = 1|\operatorname{do}(C = 0))$. If only $P(E = 1|C = 1) > P(E = 1|C = 0)$ counts, this means that there is a statistical dependency, but not implicitly a causal relationship. In literature this distinction is often referred to as correlation vs. causation [GAE06]. Besides the definitive book from Pearl there is some other literature about the detection of cause and effect relationships using probabilistic methods. A good introduction regarding probabilistic causal methods is given e. g. by Neapolitan [Nea09] or Guyon [GAE06].

In terms of process engineering the definition of the do-operator means that a system identification needs to be performed for the detection of causal dependencies. Still when testing for cause-effect relationships in data they usually need to be found based on observational data which is in conflict with the definition of the do-operator. Therefore, Pearl describes that under certain assumptions it is possible to detect cause-effect structures in data without intervening in the process. These assumptions are mainly based on the information coming from three random variables $X_i, X_j, X_k : \Omega \rightarrow \mathbb{R}$. The basic idea is based on the calculation of conditional independencies as possible types of causal structures described by three variables. Hence, table 2.1 describes all possible combinations of a graph while $X_i \perp X_j|X_k$ defines X_i being conditionally independent of X_j under the knowledge of X_k. The direction of the arrow describes the causal dependency, while $X_i \rightarrow X_j$ means that X_i has a causal influence on X_j.

Of main importance when analyzing causal structures are the three structures chain, fork and v-structure. All three consist of the same (non-causal) dependencies but in terms of cause and effect point into different directions. The

Structure	Illustration	Conditional independencies
complete connected	$X_i \leftrightarrow X_j, X_i \leftrightarrow X_k,$ $X_j \leftrightarrow X_k$	none
complete unconnected	X_i, X_j, X_k	$X_i \perp X_j, X_i \perp X_k, X_k \perp X_j$
single arrow chain	$X_i \leftrightarrow X_j, X_k$	$X_i \perp X_k, X_j \perp X_k$
chain	$X_i \to X_k \to X_j$	$X_i \perp X_j \mid X_k, X_j \perp X_i \mid X_k$
fork	$X_i \leftarrow X_k \to X_j$	$X_i \perp X_j \mid X_k, X_j \perp X_i \mid X_k$
v-structure	$X_i \to X_k \leftarrow X_j$	$X_i \perp X_j$ but $X_i \not\perp X_j \mid X_k$

Table (2.1) Possible representation of a causal graph having three variables and their corresponding conditional independencies. The double arrow $\leftrightarrow$ describes an undirected edge.

chain and the fork cover the same conditional independencies meaning that under the knowledge of X_k, X_j has no information about X_i. Having a v-structure the conditional independencies are different as without knowing X_k, X_i and X_j are independent of each other, but when knowing X_k they are not (indicated through $\not\perp$ in table 2.1).

As chain and fork have the same conditional independencies, this means that these two cannot be distinguished having observational data available and interventions need to be made. Therefore, the approach when testing for causal structures in data sets that do not contain any temporal information, the aim is primarily to find v-structures in the data. Figure 2.5 gives a closer insight by illustrating all possible graphs which can be generated for three variables and up to two edges. Structures that cover the same set of conditional independencies are also called Markov equivalent classes [Pea00], [May09]. To distinguish among Markov equivalent models, interventions need to be performed in the process [EM07] [MGS05].

Causal Markov assumption When constructing a causal graph based on probabilistic methods Neapolitan [Nea09] describes three assumptions of the underlying data set that are implicitly made. In detail these are defined as follows:

- **No hidden variables** All variables are represented in the graph. That means, all possible causes are represented as variables in the network.

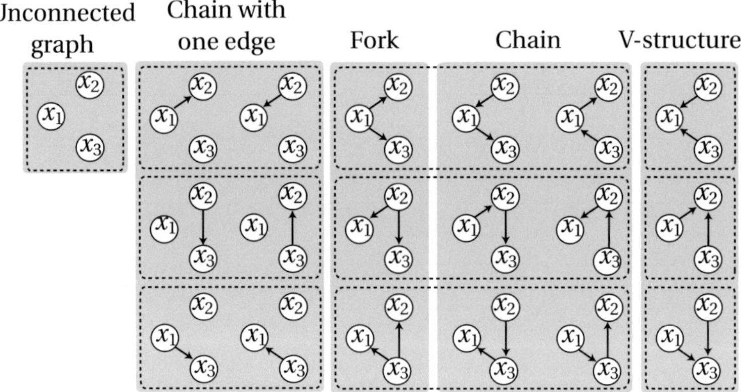

Figure (2.5) All possible causal structures for three nodes and two edges. The dotted line marks the Markov-equivalent class; the shaded gray area marks the same topology.

Especially, there is no hidden common cause which has an influence on several or even all variables.

- **No causal feedback loops** If there is a path from X_i to X_j there is no path back from X_j to X_i.

- **No selection bias** The probability distribution is obtained from a random sample of the population.

Clearly in practice all conditions can be violated. It is possible that the process consists of several control loops which cannot be detected using probabilistic measures. Furthermore, hidden variables e. g. in terms of a common disturbance that affects all process variables can have a significant impact when calculating cause-effect relationships. Therefore, it always has to be taken into account to investigate the data quality beforehand when generating a causal graph from observational data.

2.3.2 Causality and Time

In contrast to the theory of probabilistic causality the every-day life impression of causality is more connected towards time dependency. Writing this in a scientific way means that a time scale is fixed on an available data set and an event that happens at time t can only have an impact on another event that happens at time $t + dt$ but not vice versa. This concept corresponds to the definition of a causal system based on system theory which can be found e. g. in [Lun10, Föl08]. In that case, the definition says that a time-invariant system is causal if for all input signals $u_1(t) \in \mathbb{R}$ and $u_2(t) \in \mathbb{R}$ with the property $u_1(t) = u_2(t)$ for $0 \le t \le t_1$ the output signals have the property $y_1\{u_1(t)\} = y_2\{u_2(t)\}$ for $0 \le t \le t_1$ under the assumption that they have the same initial condition. In terms of causality as an additional definition it is said that in a causal system the input values for $t > t_1$ do not have an influence on output signals at time t_1 and furthermore the impulse response for $t < 0$ is zero. Additionally, following these definitions shows that for all systems for which the input is directly connected to the output, causality is given whether $u(t)$ or $y(t)$ is viewed as input, as both signals depend on each other only on the current value. To find causal dependencies in these systems probabilistic measures need to be used as temporal information is not available.

If a system has some dependency on input values from the future it is termed an acausal system [Lun10]. A possible application of an acausal filter is in terms of a postprocessing filter, as in that case the filter can extract future values from a memory buffer. Another application of these filters is in terms of image data processing as in that case data is not dependent on time but on the location, which means that in that case causality does not exist.

In the following, the two possibilities to detect cause-effect relationships in dynamic systems using time dependencies, namely delay times and dead times, are explained in detail.

Signal delay time   Delay times exist when systems react delayed on an external excitation e. g. due to their mass inertia. Systems behaving like this belong to the class of dynamic systems. This characteristic can be exploited for the detection of causal dependencies in the acquired measurement data. How these systems can be used for the detection of causal dependencies becomes obvious when writing them in form of differential equations. For example a linear-time-invariant 1st-order system with input $u(t)$ and output $y(t)$ is given as [Lun10]

$$\dot{y}(t) = \frac{K_a}{T} u(t) - \frac{1}{T} y(t), \tag{2.1}$$

with $K_a > 0$ being the amplification factor and $T > 0$ being the time constant. When having a step change in $u(t)$ the time constant describes the time the process needs to set $y(t)$ on $1 - \frac{1}{e} \approx 63.2\%$ of its final value. The information in the differential equation can be exploited for the detection of causal dependencies. This becomes clear when transforming equation 2.1 into its discretized form. How to perform a sampling of a signal is e. g. described in further detail in Åström [AW97]. The discretized system results in

$$y[kT_s + T_s] = e^{-\frac{T_s}{T}} y[k] - K_a(e^{-\frac{T_s}{T}} - 1) u[kT_s], \tag{2.2}$$

with $k = 1,\ldots,K$ samples for the time series $u[k]$ and $y[k]$ and T_s being the sampling period. As in terms of causality no information can flow from a sample point $[kT_s + T_s]$ to $[kT_s]$, equation 2.2 shows that $u[kT_s]$ and $y[kT_s]$ have a causal influence on $y[kT_s + T_s]$. In anticipation of the proposed methods in chapter 3, when investigating if u causes y, it has to be taken into account that y itself consists of causal information about its future values. Therefore, the proposed methods that use temporal information for the detection of causal influences need to calculate the information each signal has about itself before testing if a second signal has a causal influence on it.

If the system has no dynamics, meaning that $y(t) = g(u(t))$, it is not possible to detect causal behavior based on time dependency as these equations only capture instantaneous behavior. In process engineering examples of 1st-order systems are distillation columns [AMM90] or fluid tanks, in signal processing 1st-order systems describe a class of low-pass filters. Therefore, when using measurements for the calculation of the disturbance propagation path, filtering a signal in a preprocessing step has to be handled with care. This topic will be investigated in further detail in chapter 7. As benchmarks 1st-order and 2nd-order systems are used in combination with the explained disturbances defined in section 2.2.3 to test the proposed data-driven methods in chapter 3.

Signal dead time The signal dead time describes the interval a change in the input signal needs to become visible on the output of the system [Lun10]. Compared to a signal delay time this is in difference, as for the delay time changes in the input signal become visible on the output directly at the next sample. Only the characteristics of the resulting output depend on the system order, the selected sample time of the data acquisition system and the time constant of the process.

The dependency between input and output signal in the time domain for a system with dead time is therefore written as

$$y(t) = u(t - T_{\mathrm{d}}), \tag{2.3}$$

where T_{d} describes the dead time of the system. In other words, if the system consists only of a dead time element this means that $u(t)$ appears on the output $y(t)$ after T_{d}. Following the definition of temporal causality given in section 2.3.2 the information gained from the dead time can be used directly for the detection of causal dependencies. In process engineering dead times can occur in various ways.

A classic example is a fluid in a tube that passes the equipment while having a certain velocity. If a measurement device detects e. g. a disturbance at a certain point in time, this means that a similar disturbance will occur some time later in the equipment at a second device. Another example is given in chapter 6 in terms of an industrial glass forming process. As the glass flow in this plant is slow this results in large dead times for the different measurement devices.

The influence of signal dead time is given exemplarily on a tube having a certain diameter A and length l_{d} between two measurement devices. This is illustrated in figure 2.6. Assuming that the flow has a constant velocity v the dead time between the two measurement devices is calculated by dividing the volume V of the tube through its volume flow rate $\dot{V}$ [Bau05]

$$T_{\mathrm{d}} = \frac{V}{\dot{V}} = \frac{A l_d}{A v} = \frac{l_{\mathrm{d}}}{v}. \tag{2.4}$$

If the resulting value of T_d is larger than the sampling period of the data acquisition system the dead time between the two measurement devices can be used to reconstruct the fault propagation path of a disturbance. Compared to the previously explained signal delay time an important difference is that it is possible that no dynamics is contained between $u(t)$ and $y(t)$, which means that the found cause-effect relationship can also be the result from two shifted time series.

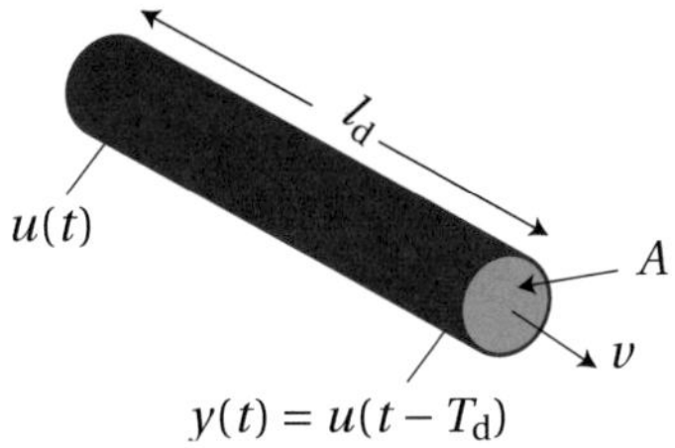

Figure (2.6) Example of how the dead time in a tube can be used for the reconstruction of the fault propagation path. A disturbance $u(t)$, measured by a process device at the entry point of the tube becomes visible at the disturbance $y(t)$ measured by a second device at the end of the tube after the dead time T_d.

2.4 Constraint-Based Learning of Causal Dependencies in Production Data

As explained in section 2.3.1 probability theory can be used to find causal structures in data when no temporal information is available. In process engineering this is the case when having for example a data set which contains quality data over several productions. Sometimes the root cause of the varying product quality is not obvious. In that case data-driven methods can be used for the detection of the main influencing parameters. In literature usually all machine learning methods that exploit statistical properties in data for the calculation of cause-effect relationships are summarized under the synonym of Bayesian Networks. There are essentially two different approaches that can be used for

the detection of causal structures in a data set. These are summarized in the
following, for more details it is referred to [Nea09] or [Mur02].

- **Search-and-score approach** Using this approach means that after defining some scoring function a greedy-search algorithm is used to find the
  graph with the maximum score. The scoring function is designed to give
  high scores to graphs that fit the data well which means that those algorithms do not test for the underlying causal relationships. In other words,
  this means that the resulting graph of the algorithm does not necessarily
  need to describe the causal structure of the system. Furthermore, there
  can be several different but highly probable structures especially when
  the sample size is small [Fri00], which can lead to falsely detected causal
  dependencies.

- **Constraint-based approach** This approach tries to find a graph which
  satisfies all the statistical constraints implied by the conditional independencies in the data. Theoretically this leads to a subset of directed
  acyclic graphs all sharing the same set of probabilistic dependency relations. Theoretically, these methods result in a partially directed graph
  which is in terms of the found cause-effect dependencies correct. Tests
  on real data sets showed that these methods can lead to problems when
  the sample size is small [TBA06] as lots of data is needed for the correct
  calculation of the statistical tests. When the number of variables in the
  data set is large these methods can lead to longer computational times
  compared to the scoring-based approaches [Mar03].

Both approaches are covered under the synonym of structure learning, which
described in further detail in [Pea00] or [Nea09]). In general structure learning
means to learn the structure of a process out of statistical data. These methods
are already applied in various fields like system biology [Fri04, EM07, SPP$^+$05],
bioinformatics [Nea09] or psychology [Med06, Gly01]. In process engineering
examples that use Bayesian Networks can be found in condition monitoring
tasks [LFR06] or as decision support systems [WMD02]. In the present work
it is demonstrated how structure-learning can be used for the detection of
the root cause of a disturbance using the constraint-based structure learning

approach. Therefore, an algorithm originally proposed by Peter Glymour and Clark Spirtes [SGS00], called the PC-algorithm (named after the forenames Peter and Clark), is applied. This method is used to analyze production data coming from a simulated continuous stirred tank reactor. The method is chosen as it theoretically detects all possible causal dependencies in a data set that can be found with a purely data-driven approach. Nevertheless, for this approach the focus is less on the method, as possibly other methods can be used as well. The used concept for the detection of causal structures is more important when having production data at hand.

2.4.1 PC-Algorithm

As mentioned, the PC-algorithm exploits constraints in the data in terms of conditional independency statements to find cause-effect relationships. In practice this means that the conditional independencies are found by performing statistical tests on the available data set. In so doing the algorithm works in two stages.

First it starts with a completely connected graph and calculates the undirected graph structure by removing edges between (conditionally) independent variables. The resulting graph is called the skeleton.

In a second step the algorithm orients edges by first searching for possible v-structures in the data and afterwards by following some logic rules to direct chains and forks. By definition, it is assumed that the Causal Markov Assumptions as explained in section 2.3.1 hold for the data set. For the explanation of the algorithm some notifications need to be made.

Formally a graph $\mathbb{G} = (V, E)$ consists of a set of nodes $V = \{X_1, X_2, \ldots, X_k\}$ and a set of edges $E \subseteq V \times V$. In common terminology $X_i \rightarrow X_j$ means that X_i is a parent of X_j and X_j is a child of X_i. Following the definition of Pearl this also describes a causal dependency. An existing edge $X_i \leftrightarrow X_j$ means that X_i and X_j are adjacent and the set being adjacent to X_i is described as $\mathbf{Adj}_{X_i}$. $|\mathbf{Adj}_{X_i}|$ is defined as the number of nodes of all variables being adjacent to X_i. If $\mathbb{G}$ contains only directed edges and no cycles it is called a directed acyclic graph (DAG). A partially directed acyclic graph (PDAG), which is the result of the PC-algorithm, contains directed and undirected edges and represents the Markov equivalence class of a DAG.

In terms of causality it is defined that X_i is a direct cause of X_j if $X_i \rightarrow X_j$ exists and an indirect cause if X_i has an influence on X_j over some other variable e.g. through $X_i \rightarrow X_k \rightarrow X_j$. Furthermore, it is said that X_j is conditionally independent of X_i if it has only an indirect causal dependency.

Formally the variables X_i with $p \in \{1,\dots,D_p\}$ and X_j with $q \in \{1,\dots,D_q\}$ are defined to be conditionally independent given a third variable X_k with $r \in \{1,\dots,D_r\}$ if

$$P(X_i = p | X_j = q, X_k = r) = P(X_i = p | X_k = r). \tag{2.5}$$

Testing the variables X_i and X_j for conditional independency against the remaining variables $V \setminus \{X_i, X_j\}$ in the graph is the core of the PC-algorithm and the detection of v-structures. For the test the G^2 statistic is used which is described in section 2.4.2.

Detecting the skeleton The algorithm starts with a completely connected undirected graph $\mathbb{G} = (V, E)$ and iteratively removes edges by performing (conditional) independency tests. For each variable $X_i \in V$ an independency test against any other variable $X_j \in V \setminus X_i$ is made. If they are independent, the edge between X_i and X_j is removed and a variable m which counts the size of the subset of variables is set to 1.

For each variable $X_i \in V$ the adjacent nodes $\mathbf{Adj}_{X_i}$ are selected. Next it is tested, if X_i is conditionally independent of X_j given any subset S of $\mathbf{Adj}_{X_i} \setminus X_j$ of size m. If $X_i \perp X_j | S$ counts, the link between X_i and X_j is removed and the separating set S is stored. Next the set S is added to the variable set $S_{X_i X_j}$ as this is needed for the orientation of the edges. Finally m is increased by one and the algorithms stops as m reaches $|\mathbf{Adj}_{X_i}|$.

Directing the edges As a next step the v-structures are directed and therefore the uncoupled combinations $X_i - X_k - X_j$ need to be tested for conditional independency. This is done by checking if the variable X_k is in the set $S_{X_i X_j}$ as for a v-structure $X_i \perp X_j$ and $X_i \not\perp X_j | X_k$ counts (see section 2.3.1). This means if X_k is not in the set $\boldsymbol{S}_{X_i X_j}$, a v-structure is detected and the edges are oriented as $X_i \rightarrow X_k \leftarrow X_j$.

Additional edges can be directed based on further rules e. g. that no cycles are allowed and that no further v-structures are possible in the final graph. For a detailed explanation on how to direct these edges it is referred to [Pea00] or [Nea09]. In detail these are: Orient $X_j \rightarrow X_k$ if $X_i \rightarrow X_j$ exists and no connection from $X_i \rightarrow X_k$; orient $X_i \rightarrow X_j$ if $X_i \rightarrow X_k \rightarrow X_j$ exists; orient $X_i \rightarrow X_j$ if $X_i - X_k \rightarrow X_j$ and $X_i - X_l \rightarrow X_j$ exist while X_k, X_l are not connected; orient $X_i \rightarrow X_j$ if $X_i - X_k \rightarrow X_l$ and $X_k \rightarrow X_l \rightarrow X_j$ exist while X_k, X_j are not connected. Further information about the algorithm can be found e. g. in Pearl [Pea00] or Neapolitan [Nea09].

2.4.2 Test Statistics

To test a pair of variables X_i, X_j for conditional independency against a subset of variables $\boldsymbol{X_k} = \{X_1, \ldots, X_k\}$, described as $X_i \perp X_j | \boldsymbol{X_k}$ the G^2 test proposed by Tsamardinos [TBA06] is used. This is a hypothesis test which has the null hypothesis that the two variables X_i, X_j are conditional independent from each other. The test operates with quantified data and the G^2 statistic is in that case calculated as

$$G^2 = 2 \sum_{r=1}^{D_r} \sum_{q=1}^{D_q} \sum_{p=1}^{D_p} S^{pqr}_{X_i X_j \boldsymbol{X_k}} \ln \frac{S^{pqr}_{X_i X_j \boldsymbol{X_k}} S^{r}_{\boldsymbol{X_k}}}{S^{pr}_{X_i \boldsymbol{X_k}} S^{qr}_{X_j \boldsymbol{X_k}}}. \tag{2.6}$$

In that case $S^{pqr}_{X_i X_j \boldsymbol{X_k}}$ describes the number of times in which the variable X_i has the value p, the variable X_j has the value q and the variable X_k has the value r. The variables are defined the same way for $S^{r}_{\boldsymbol{X_k}}, S^{pr}_{X_i \boldsymbol{X_k}}, S^{qr}_{X_j \boldsymbol{X_k}}$. The resulting value for G^2 can be transformed into a p-value from the χ^2 distribution with appropriate degrees of freedom. If the p-value is below a defined significance level α, which is set to 0.05, the null hypothesis holds meaning that X_i and X_j are defined conditionally independent. According to [TBA06] the degrees of freedom df can be calculated as

$$df = (D_p - 1)(D_q - 1) \prod_{l:X_l \in \boldsymbol{X_k}}^{\boldsymbol{X_k}} D_l, \tag{2.7}$$

where D_p, D_q and D_l define the binning size of each variable.

In addition there exist several other tests for the detection of conditional independencies. The most common ones are the Pearson's $\mathcal{X}^2$ test and an information-theoretic distance measure based on mutual information. A good overview is given in [Mar03] and [SB11]. In the present work these two tests will not be considered in further detail.

2.4.3 Continuous Stirred Tank Reactor

This section illustrates how the concept of probabilistic causality can be used for process optimization and the detection of the root cause of a disturbance in production data. This is done by using the introduced PC-algorithm, but could also be performed by other structure learning algorithms. The data set is generated from a continuous stirred tank reactor (cstr) running in feed-forward control. The cstr model is derived from a real plant and has been presented by Rawling [TR02] in 2002. The underlying chemical reaction scheme consists of two irreversible follow-up reactions where an educt A reacts to an intermediate product B and this again reacts to the resulting product C. The reactants are dissolved in the fluid and can be measured in terms of the three concentrations c_A, c_B, c_C at the outlet of the cstr. During a run, the cstr model is continuously filled with the reactant which is dissolved in the concentration c_{in} having the temperature ϑ_{fl}. Depending on ϑ_{fl} a larger or lower quantity of the products B and C are produced as it has an exponential impact on the reaction. A schematic drawing of the process is given in figure 2.7 on the left hand side. The underlying differential equations of the cstr are described as

$$\dot{c}_A(t) = \frac{F}{V}(c_{in}(t) - c_A(t)) - k_1 c_A(t) e^{-E_1/(R\vartheta_{fl})}, \tag{2.8}$$

$$\dot{c}_B(t) = k_1 c_C(t) e^{-E_1/(R\vartheta_{fl})} - k_2 c_B(t) e^{-E_2/(R\vartheta_{fl})} - \frac{F}{V} c_B(t), \tag{2.9}$$

$$\dot{c}_C(t) = k_2 c_B(t) e^{-E_2/(R\vartheta_{fl})} - \frac{F}{V} c_C(t). \tag{2.10}$$

The parameter V describes the volume of the cstr and F is the volume flow rate which is kept constant during the production. The parameters k_1 and k_2

are empirical parameters which are pre-exponential factors and describe the relationship between the temperature and the speed of the chemical reaction. E_1 and E_2 describe the activation energy of the reactants and R describes the universal gas constant. The resulting causal relationships can be deduced from the differential equations and are shown in figure 2.7 on the right hand side in terms of a directed graph. The values of the cstr model are shown in Table 2.2 and are equivalent to those presented by Rawling.

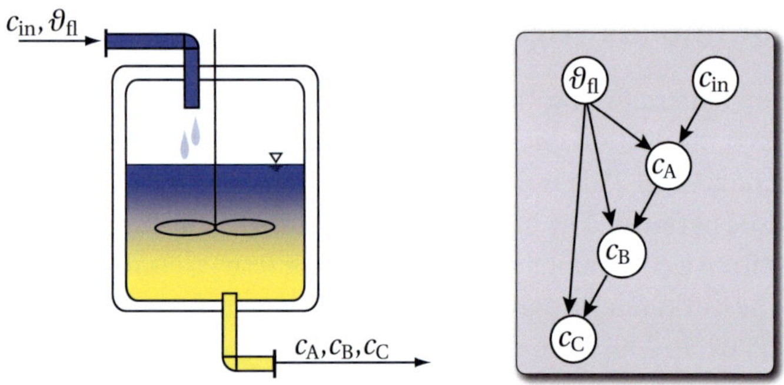

Figure (2.7) Illustration of used continuous stirred tank reactor. The left plot sketches the plant, the right plot illustrates the ground truth of the causal structure derived from the differential equations.

Parameter	Value	Unit
F	100	L/min
V	100	L
k_1	7.2×10^{10}	1/min
k_2	5.2×10^{10}	1/min
E_1/R	8750	K
E_2/R	9750	K

Table (2.2) Selected model parameters for the continuous stirred tank reactor taken from [TR02].

The set-points of the two input variables are chosen as $\vartheta_{fl,OP} = 350\,$K and $c_{in,OP} = 1\,$mol/L. If there is no additive noise superposed on the measurements the other process variables adopt in stationary phase the values $c_{A,OP} = 0.5\,$mol/L, $c_{B,OP} = 0.48\,$mol/L and $c_{C,OP} = 0.025\,$mol/L. For further information how to model chemical reactions it is referred to [Ari99] or [Ing07].

Generation of the data set   For the generation of the data set the set-point $\vartheta_{\mathrm{fl,OP}}$ is superposed with the noise variable $v_\vartheta \sim \mathcal{N}(0,7\,\mathrm{K}^2)$ and $c_{\mathrm{in,OP}}$ is superposed with the noise variable $v_{c_{\mathrm{in}}} \sim \mathcal{N}(0,0.15\,(\mathrm{mol/L})^2)$ meaning that concentration and temperature differ on a run to run basis. Due to the fact that the process runs in feed-forward control each concentration finishes on a different stationary value after start-up. The resulting stationary values are stored in a database on a run-by-run basis and are used for the detection of causal dependencies. To avoid numerical issues $K = 10000$ start-up runs are simulated. To get a better view of the system dynamics, figure 2.8 shows the resulting mean value μ and the standard deviation σ for the simulated start-up runs of the cstr for each process variable. The purpose is to reconstruct the causal structure as shown in figure 2.7 and to detect the process variables being on top of the causal chain as they have the main influence on the varying concentration of the resulting products. Regarding industrial practice this means that one wants to find process parameters that have an early impact on the later resulting product quality and the plant can be optimized by intervening in an early stage of the chemical reaction. The variations of ϑ_{fl} and c_{in} can also be interpreted as disturbances acting on the system.

Results   For the calculation of the G^2 test, the data of each process variable is divided into $D = 10$ equally sized intervals with the condition that the number of data points in each interval is constant. Afterwards, every data point is replaced by the label of its interval.

The resulting graphs are outlined in figure 2.9 in terms of the initial complete graph, the detected skeleton and the final PDAG. Comparing the skeleton with the complete graph indicates that it only differs in the missing link between ϑ_{fl} and c_{in}. When analyzing the resulting PDAG the two input variables ϑ_{fl} and c_{in} are correctly detected as root cause of the disturbance meaning that for process optimization these variables are the first to be changed if one wants to increase the resulting concentrations c_B and c_C of the products. Furthermore, the algorithm finds the causal dependency $c_A \rightarrow c_B$ and $c_A \rightarrow c_C$ which reveals

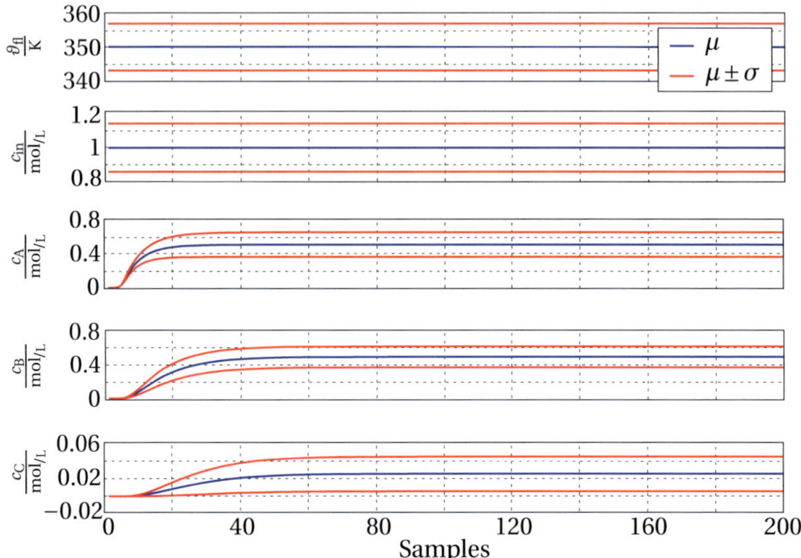

Figure (2.8) Start-up behavior of the simulated cstr. As the process runs in feed-forward control with variations in ϑ_{fl} and c_{in} the resulting stationary values of the products differ. This information can be used to detect causal dependencies in the data and identify ϑ_{fl} and c_{in} as root causes.

details of the elementary chemical reaction. The second follow-up reaction $c_B \rightarrow c_C$ could not be detected as the graph indicates an intercausal relationship meaning that according to the PC-algorithm c_B has a causal influence on c_C and c_C back on c_B. The reason why the reaction could not be detected is that the disturbance implied in c_{in} is already filtered two times through the other reaction schemes which reduces the variance in the data set for the detection of $c_B \rightarrow c_C$. Using the G^2 test the results indicate that the v-structures can be detected in the data set, but it is not possible to distinguish between direct and indirect causal dependencies. This becomes obvious as the edges $c_{\mathrm{in}} \rightarrow c_B$, $c_{\mathrm{in}} \rightarrow c_C$ and $c_A \rightarrow c_C$ were all selected by the algorithm. Still, this can be tolerated as they all point into the direction of the causal flow meaning that ϑ_{fl} and c_{in} are correctly found as root causes.

To summarize the results, the generated data set from the cstr shows that it is possible to detect causal structures in production data and to gain new process knowledge for process optimization by localizing the root cause of the varying product quality. In the present work, the PC-algorithm, which theoretically detects all causal dependencies up to its Markov equivalent class, was used to determine if causal structures can be found in production data based on the concept of probabilistic causality. There is a large variety of further scoring- and constraint-based structure learning algorithms available which all have their benefits and drawbacks. Since the main focus in this work is on fault localization implying temporal information in data the discussion about the detection of causal structures using probabilistic methods is carried out in chapter 8 as a possible future research.

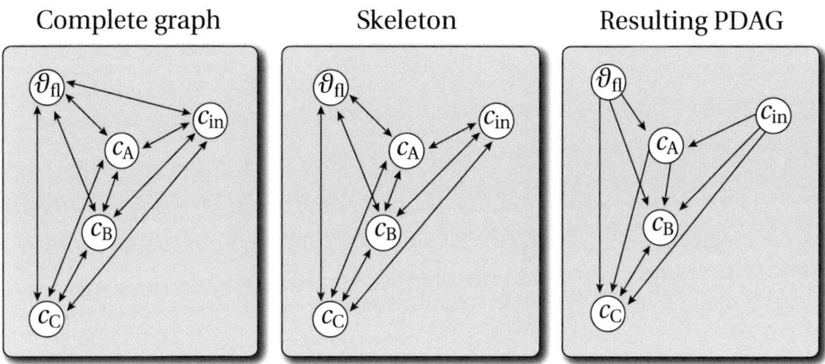

Figure (2.9) Results of the PC-algorithm for the detection of causal structures using data from the continuous stirred tank reactor. Edges with two arrowheads indicate an intercausal dependency meaning that both variables have a causal influence onto each other. ϑ_{fl} and c_{in} are correctly set on top of the reaction chain.

3

Data-Driven Methods for Fault Localization

This chapter starts by classifying the suggested methods regarding their theoretical framework and introduces several benchmark data sets for testing them. Thereafter several visualization techniques for the representation of the found causal dependencies are reviewed and those used in this thesis are explained in further detail. The main part of this chapter explains the algorithmic details of the proposed methods. Finally, each method is tested on the proposed benchmark data sets.

3.1 Method Classification

The possible methods used for fault localization and the calculation of the disturbance propagation paths can be separated into three different classes represented in figure 3.1, namely machine learning, time-frequency analysis and statistical signal processing. In this case the classification is not distinctive, meaning that some of the methods could be allocated into several classes. The methods marked in bold are investigated in further detail in this thesis, while as illustrated in figure 3.1, there are other methods possible for the detection of cause-effect relationships and the reconstruction of disturbance propagation paths. An explanation of the characteristics of these classes is given below.

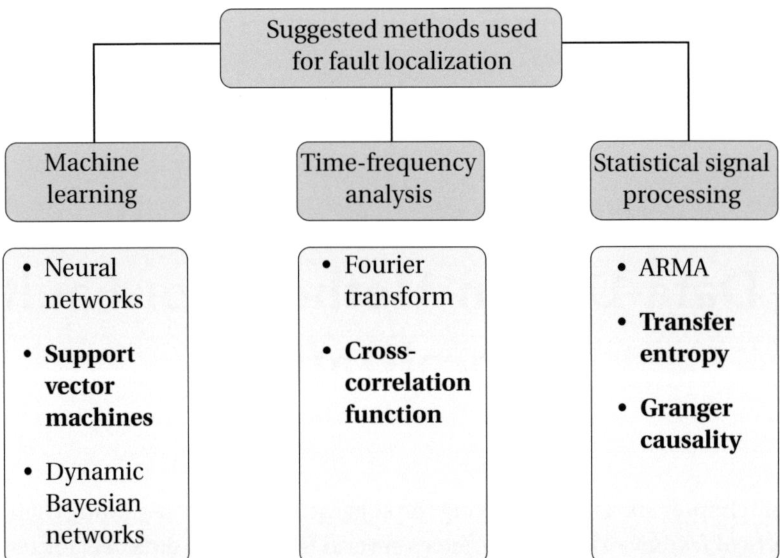

Figure (3.1) Survey of possible methods used for the detection of causal dependencies in process data. The methods marked in bold are used in the present work for fault localization.

- **Machine learning** Machine learning is most often used when performing pattern recognition in data sets and can be seen as a part of artificial intelligence. The concept of these methods is that computers can learn a certain behavior by using empirical data. After having a certain data set used for training the machine learning algorithm can be utilized for classification or regression tasks of new input data. There exists a wide variety of different machine learning algorithms used for pattern recognition. A good overview is given e. g. in [WF05] or [WKRQ$^+$07] but there is also lots of other additional recent literature available. In the present work Support vector machines (SVM), described in further detail in section 3.7, are used for detecting cause-effect relationships in the data.

- **Time-frequency analysis** In the field of electrical engineering time-frequency analysis combines classical methods that are used for analyzing continuous or discrete time signals. The concept when using these methods is that signals in their time and frequency representation are tightly connected and can be understood better when being analyzed in different domains. The most common method used when performing a time-frequency analysis is the calculation of its Fourier transform. The Fourier transform transfers a signal which is dependent of time into a new function which then uses as argument the frequency in hertz (or radians per second). In the present work, from this area of data analysis, the cross-correlation function is applied (section 3.4) as a causal measure. The concept of this method is to calculate the linear similarity of two signals while delaying one signal against the other. The time lag and the amplitude of the resulting cross-correlation coefficient is then used to give indications of possible causal dependencies in the data. According to the Wiener-Khinchin theorem [Joh93], the cross-correlation function of two signals in the Fourier transform results in the cross power spectrum. Using the cross power spectrum for causal analysis will not be regarded in further detail in this thesis. More explanations about time-frequency analysis is given e. g. in [Joh93] or [Ise06].

- **Statistical signal processing** Statistical signal processing treats signals as stochastic processes and uses their statistical properties for analysis. Classical methods are autoregressive moving average models (ARMA) or the calculation of entropic measures in information theory. Examples of processes that are modeled as a stochastic time series are the stock market [KV11] or the Brownian motion of molecules [DOT03]. In the present work transfer entropy (section 3.5) and Granger causality (section 3.6) are used to detect causal dependencies in data. The main concept of Granger causality is grounded on the estimation of vector auutoregressive models and their application as a one-step-ahead predictor; the idea of transfer entropy is based on the comparison of transition probabilities calculated from the investigated time sequences. An introduction towards statistical signal processing is given e. g. in [Gra10], an overview of classic methods mainly used for system identification can be found in [Joh93].

3.2 Benchmark Data Sets for Fault Localization

After introducing the concept of using statistical relations and temporal information for the detection of cause-effect dependencies in data several benchmarks need to be defined to evaluate the characteristics of the methods. In that case the aim is two-folded. First it has to be checked which method reliably detects the underlying causal structure and second how the method will behave if the dependency is not correctly found. Hence, the benchmarks mainly consist of 1st-order systems that differ concerning their structural characteristics in terms of time delays, nonlinearities, feed-through or feedback loops. Additionally, the behavior of the methods using the explained disturbances in section 2.2.3 is investigated. In detail the disturbances are white noise, colored noise and a step function with superposed white noise. The sinusoidal input signal will be investigated separately in chapter 7. To study the characteristics of these methods on a higher order process the benchmarks comprises a vibratory 2nd-order system as well. As an example of a MIMO system the behavior of the methods is investigated on the model of a continuous stirred tank reactor which was already introduced in section 2.4.3.

Except the MIMO system, all benchmarks use an amplification factor $K_a = 1$ and a time constant $T = 0.5\,$s. The input signal $u(t) \in \mathbb{R}$ consists of zero mean white noise with unit variance $\mathcal{N}(0, 1)$. The sample time is selected as $T_s = 0.1\,$s. In addition, a separate signal $z(t) \in \mathbb{R}$, consisting of colored noise, is generated from another source $v(t) \in \mathbb{R}$ having white noise with $\mathcal{N}(0, 1)$ and a 1st-order system. This is illustrated in figure 3.3. This variable has no influence on the other variables in the benchmark data set but used for the calculation of the causal matrix and needs to be sorted out by the methods.

To avoid numerical issues each data set consists of $K = 10000$ samples. The used benchmarks are illustrated in figure 3.2 and are explained below:

- **Base configuration** This is the most basic configuration, containing a standard 1st-order system.

- **Squaring device** The output signal $y(t)$ is squared. The main purpose is to determine which methods can detect nonlinear causal relationships

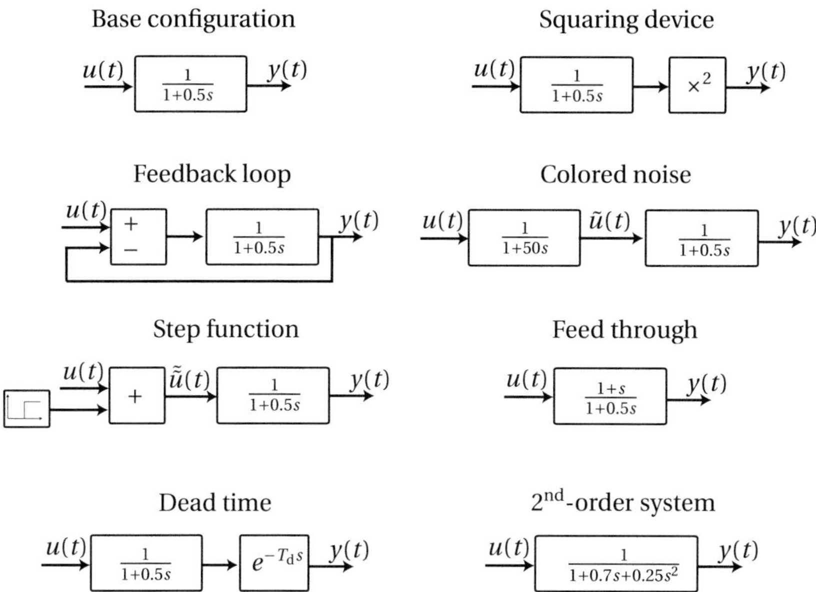

Figure (3.2) Utilized benchmarks for the detection of causal relationships in dynamic systems. The benchmarks consist of different noise scenarios as well as different process characteristics. In all benchmarks the s represents the Laplace transformation of the impulse responses from LTI systems.

in the data set. A squaring device is used since this describes an even function. This prevents that the causal relationships can be estimated by using a linear approach, which would be the case when using uneven functions.

- **Feedback loop** This benchmark consists of a negative feedback loop. In that case, the output signal $y(t)$ is substracted from the input signal $u(t)$. The amplification of the feedback loop is set to one.

- **Colored noise** The signal $u(t)$ is low-pass filtered to generate an input signal $\tilde{u}(t)$ with a limited bandwidth as input for the 1st-order system. This signal is subsequently used for the detection of the causal dependency

Signal from separate process

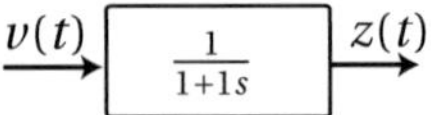

Figure (3.3) Generated signal $z(t)$ coming from a second process. The proposed methods need to sort out this signal in the benchmark data set, as otherwise a wrong causal dependency would be detected.

of the 1st-order system. The outcome of this benchmark is of special importance as usually disturbances in a plant are narrowbanded.

- **Step function** After half the simulation time a step with the amplitude $K_{\text{step}} = 10$ is superposed on the input signal $u(t)$. This benchmark is used to determine how the different methods behave when having a non-stationarity in the disturbance.

- **Feed through** The 1st-order system is changed as an additional lead element is used to generate a system with a feed-through. This means that the output signal $y(t)$ consists of a direct and a time delayed impact from $u(t)$. As the direct part does not contain any information useful to detect causality the methods need to filter it out to detect the underlying cause-effect relationship.

- **Dead time** The system consists of a dead time which is set to $T_{\text{d}} = 2\,\text{s}$. As the sampling time is set to $T_{\text{s}} = 0.1\,\text{s}$ this means that a change on $u(t)$ has an impact on the output after 21 samples.

- **2nd-order system** In this benchmark the behavior of the different methods is tested on a mass-spring-damper system. As for the 1st-order system the time constant is set to $T = 0.5$, the dimensionless damping ratio is selected to be $\zeta = 0.7$.

3.3 Calculation of the Causal Matrix

When calculating causal dependencies in data an interpretable representation
of the resulting cause-effect relationships is of crucial importance. Therefore, to
represent the complete causal interaction of all investigated process variables
a matrix Q, as shown in equation 3.1, is generated. This matrix consists of
the r investigated process variables $X_i = \{X_1...X_r\}$ and has as values the causal
strengths $q_{X_i \to X_j} \in \{0,...,1\}$. The variable $q_{X_i \to X_j}$ is a continuous heuristic measure which describes the causal impact one variable has onto another and
increases monotholically with an increasing causal strength. For all methods
investigated in this thesis values close to 0 describe weak causal strengths, values close to 1 describe strong ones. Furthermore, the values on the diagonal
axis are not calculated. To get the full information of all causal dependencies
from the process, $r(r-1)$ values need to be calculated.

$$Q := \begin{bmatrix} - & q_{X_2 \to X_1} & \cdots & q_{X_r \to X_1} \\ q_{X_1 \to X_2} & - & \cdots & q_{X_r \to X_2} \\ \vdots & \vdots & \ddots & \vdots \\ q_{X_1 \to X_r} & q_{X_2 \to X_r} & \cdots & - \end{bmatrix}. \tag{3.1}$$

Besides the suggested data-driven approaches in this work, there are several
other ways to generate causal matrices. If the physical model of the process
is known the underlying differential equations can be used for the generation
of the causal matrix as briefly described in section 2.3.2. Another way for
constructing these matrices is to use expert knowledge, coming from a process
engineer or the plant schematic. Since in this work, there are several data-
driven methods used on one data set and as each method will have as outcome
one causal matrix, the resulting matrices need to be represented appropriately.
The following section illustrates how this can be done in a suitable manner.
Combining these matrices to one resulting causal matrix will be explained in
detail in section 4.2.

Root cause priority list   This list contains a ranking of the analyzed process variables with regard of their possibility of being the actual root cause. As a consequence a value, defined as RC, is given to each variable which is calculated from the causal matrix Q. This is done for all process variables $n = 1...r$ by summing up the causal influence from one variable onto the other variables defined as

$$\text{RC}_n := \sum_{i=1, i \neq n}^{r} q_{X_n \to X_i}. \tag{3.2}$$

The resulting variable having the maximum value of RC is ranked first, meaning that this variable is most possible to be the root cause, followed by the other variables ranked in descending order. Table 3.1 outlines the representation of the root cause priority list as used in the thesis. In section 3.3.2 it is described how this hit list can be used for the generation of a receiver operating characteristic to evaluate the outcome of the methods.

Rank	Process variable	RC
1	X_n	$\sum_{i=1, i \neq n}^{r} q_{X_n \to X_i}$
$\vdots$	$\vdots$	$\vdots$
r	X_k	$\sum_{i=1, i \neq k}^{r} q_{X_k \to X_i}$

Table (3.1) Root cause priority list generated from the causal matrix. The variable ranked first is selected as being most probable the root cause of the disturbance.

3.3.1 Visualization

For the visualization of the causal matrices several representations are used in this thesis as each type of visualization needs to fulfill several tasks. Of main importance is that the representation needs to give a qualitative impression which can be quickly understood by a process engineer, especially the root cause has to be clear as this localizes the position of the fault that leads to the disturbances. Furthermore, the disturbance propagation path has to become obvious out of the visualization. To have a quantitative measure, the causal strength of each relationship needs to be illustrated and if several methods are used on the same data set an appropriate representation for the results of the different methods needs to be found. Several techniques have been already developed which deal with this problem.

Seth [Set10] suggests using circular directional charts. In that case all process variables are arranged in a circle while being separated by an angle. A threshold is defined and all values above it are represented as a directed edge from one variable to another. For the representation of a disturbance propagation path this method can become ambiguous when a large number of variables is used, as edges can cross each other.

As another visualization technique Bauer [Bau05] proposes to use bubble charts, where strong causal dependencies are represented as large bubbles and weak dependencies as small bubbles. The concept of using bubble charts has been transferred from economics where those charts are used to represent a function of two inputs and one output. In her work she uses this representation in addition to directed graphs which will be explained at the end of this section. Eaton [EM07] suggests using heat maps in which values between zero and one are translated into different colors. He uses the representation for the illustration of causal dependencies for microcellular data. Still heat maps are hard to be understood intuitively and it is complicated to detect directly the root cause in this type of representation. For illustration, figure 3.4 shows all three types of visualization on an arbitrary causal matrix. One drawback of all these types of representation is that only one causal matrix yielding from one method can be represented at a time. This makes a comparison of the resulting causal matrices from more than one method infeasible.

As in this work several methods are developed and compared to each other, different types of visualization are needed. These types of visualization are represented in this chapter in terms of a partially directed graph, doughnut chart and as a bar chart. Each causal matrix can be represented in these three different ways explained below.

Partially directed graph   Partially directed graphs, which are also called flow charts, are intuitively understood when used for representing the information of the calculated causal matrix. In these graphs, each process variable is represented by a node and each edge represents the causal dependency from one

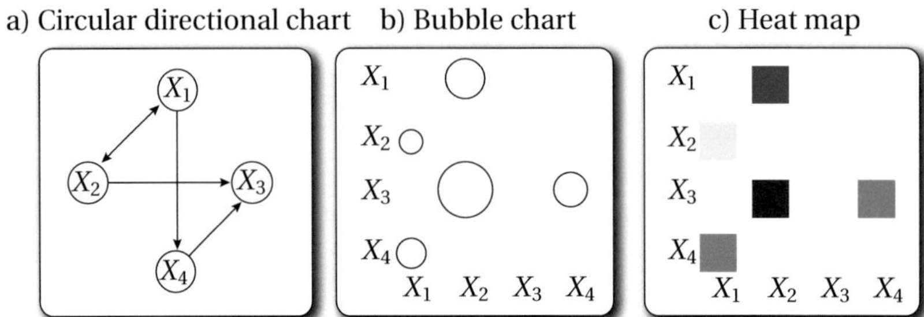

Figure (3.4) Survey of several possibilities on how to visualize a causal matrix Q. a) shows the causal matrix in terms of a circular directional chart, b) as a bubble chart and c) in terms of a heat map, while all three representations show the same causal matrix. With the circular directional chart it is not possible to visualize the causal strength, the bubble chart shows it in form of the size of its circles, the heat map in form of the selected color.

variable on another. In partially directed graphs it is possible that several edges point onto one node or that several edges leave one node. The most basic case in terms of the 1st-order system (see equation 2.4) is illustrated in figure 3.5 in the lower plot. In that case the underlying causal structure is a directed edge from $u(t)$ to $y(t)$. The main purpose of this representation is to give the process engineer a fast overview of the disturbance propagation path, while the root cause is the first variable of the chain. Since, except the cross-correlation function, all proposed methods can detect intercausal dependencies, which means that two variables can have a causal influence onto each other, edges in the graph can consist of two arrowheads pointing on the two different process variables. As all developed methods calculate a causal impact yielding a value from zero to one, the size of the arrowhead is used to indicate the strength of the causal dependency. Hence, the main found causal dependencies are illustrated with large arrowheads and weak causal dependencies with small ones. The main drawback of the representation is that only one matrix, e. g. the resulting combined causal matrix using all methods can be represented which makes this type of representation not usable for a comparison of the methods.

Doughnut chart These graphs are circular charts which are divided into several sectors, while having a blank center. The arc length of each sector is in that case proportional to the quantity each sector represents in the doughnut chart. To represent the causal matrices, the quantity of each sector results from the calculated entries in Q of the tested methods plus one blank sector. The blank sector is needed, as the calculated causal strengths are usually smaller than one, which means that the circular chart is not completely filled. The value written in the middle of the doughnut represents the calculated combined value of all used methods. The values passing the defined threshold are visualized in the partially directed graph as well.

As an example, figure 3.5 represents the 1st-order system in which two arbitrary methods are tested on the data set while the first method detects a causal strength with value of one and a second method detectes a causal strength with a value of 0.5. Doughnut charts are used in section 2.4 when testing the methods on the benchmark data sets, on the laboratory plant in chapter 5 and the industrial glass forming process in chapter 6.

Bar chart Bar charts represent values in form of rectangular bars while having their length proportional to the represented value. In this thesis bar charts are especially used for a better comparison of the different methods used on the same data set. Compared to the doughnut charts where the sum of the causal strengths of all methods for one process pair needs to be normalized to fit $360°$, this is not needed for the bar chart, which makes the comparison of the methods easier. Furthermore, this visualization avoids a further drawback of a doughnut chart as the different sections are hard to compare because they are bent. Figure 3.5 shows a bar chart on the right hand side for the 1st-order system. This type of visualization is used when the different methods are tested on the laboratory plant in chapter 5, on the industrial glass forming process in chapter 6 and when investigating the impact of sample time, compression and oscillations in chapter 7.

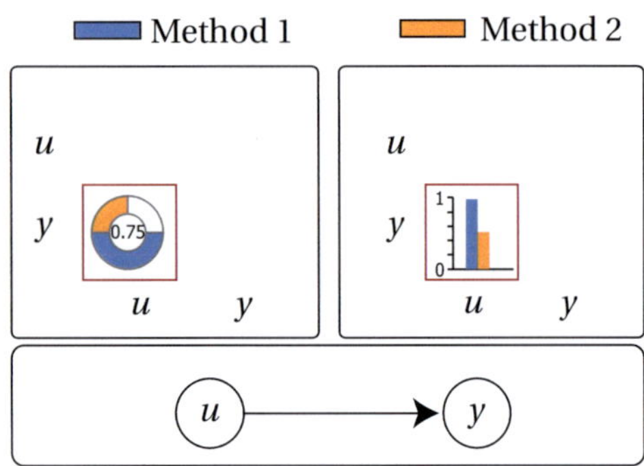

Figure (3.5) Exemplary visualizations of the causal dependency of the 1st-order system from equation 3.2 for two arbitrary methods. The dependency is given in three shapes, namely as a doughnut chart (left), a bar chart (right) and as a partially directed graph (bottom). The red square illustrates the expected dependency.

3.3.2 Receiver Operating Characteristic

The receiver operating characteristic (ROC) in its original form is used to illustrate the performance of a binary classifier while varying one parameter. Each time the parameter is varied the true positive rate of the classifier is calculated and plotted against the false negative rate. When plotting the ROC of a classifier the curve gives advices how to select the varied parameter to give the intended classification results and illustrates the overall performance of the classifier. ROC analysis was first used for the analysis of radar signals and is explained in many statistical textbooks, e. g. [Hal08] or [PB03].

In the present work, an adaptation of the original ROC curve is used to evaluate the performance of the proposed methods. Therefore, the root cause priority list as described in section 3.3 is calculated several times while using each time a different disturbed data set coming from the same fault. From this set of priority lists, the probability of the detection of the root cause variable is calculated by checking if the corresponding variable is in the first n ranked variables. The result is compared to a random selection of the variables. An example plot of a ROC is given in figure 3.6. The diagonal line in blue represents the results for the detection of the fault performing a random selection of the process variables. Hence, the line is defined as $f(i) = \frac{i}{r}$ where i is the number of selected variables. The curves in yellow, red and green show possible outcomes when calculating the probabilities of detecting the root cause. The boundary values of the curve correspond to those of the random selection. Selecting zero variables leads to the trivial result that the probability of finding the root cause is zero. Investigating all variables, the detection probability corresponds to one. The theoretical optimum is achieved, when the probability is one for the detection of the root cause, when investigating only the first variable. If the curve is close to the diagonal line the found causal matrix is no better than performing a random selection of the process variables. A ROC analysis is performed in chapter 4.3 on the continuous stirred tank reactor and in chapter 5 on the laboratory plant.

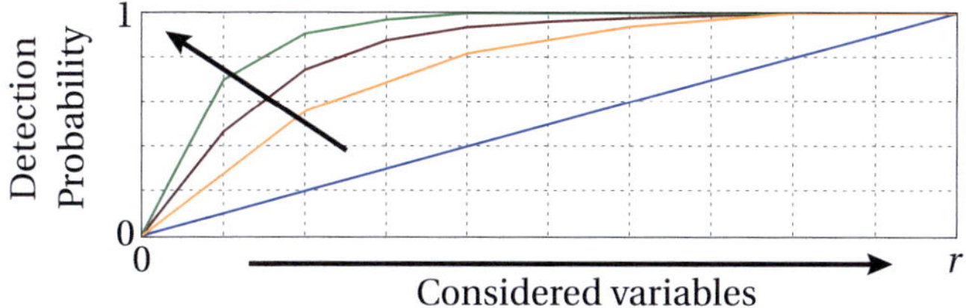

Figure (3.6) Illustration of the ROC. The different curves in the plot represent the probability of the detection of the root cause variable depending on the number of investigated variables. Therefore, the major goal is to have a ROC which is as close as possible to the upper left corner.

3.4 Cross-Correlation Function

The cross-correlation function (CCF), which is e. g. introduced in detail by [Joh93], measures the linear similarity of two equidistant sampled time series $u[k] \in \mathbb{R}$ and $y[k] \in \mathbb{R}$ with $k = 1,\ldots,K$ samples that are delayed from each other in time by a constant delay λ. By selecting the point of maximum correlation of the two time series, the corresponding value of λ is used to indicate a causal dependency in terms of a time-shifted correlation of the two signals and is finally applied to calculate the causal matrix.

Since the CCF is based on correlation it is restricted to linear systems, but it has the advantage that its outcome is easily understood and, as outlined in [Bau05], the CCF is quite tolerant to additive noise in the data. The CCF is therefore estimated as

$$
\hat{c}_{uy}[\lambda] = \frac{1}{K} \frac{\sum_{k=1}^{K-\lambda} (u[k] - \hat{\mu}_u)(y[k+\lambda] - \hat{\mu}_y)}{\sqrt{\sum_{k=1}^{K-\lambda} (u[k] - \hat{\mu}_u)^2 \sum_{k=1}^{K-\lambda} (y[k] - \hat{\mu}_y)^2}},
\tag{3.3}
$$

with $\lambda \in \{1 - K, 2 - K, \ldots, K - 2, K - 1\}$ and

$$
\hat{\mu}_u = \frac{1}{K} \sum_{k=1}^{K} u[k], \qquad \hat{\mu}_y = \frac{1}{K} \sum_{k=1}^{K} y[k].
\tag{3.4}
$$

As the denominator describes the standard deviations of $u[k]$ and $y[k]$ the resulting CCF is standardized on $\hat{c}_{uy}[\lambda] \in [-1, 1]$. If $\max|\hat{c}_{uy}[\lambda]| = 1$ this means that $u[k]$ and $y[k]$ are perfectly correlated at a shift λ, while values close to zero indicate that no correlation exists between the signals. Formally it is stated that under the assumption that $u[k]$ is a realization of a stationary white noise process and that the connecting system of $u[k]$ and $y[k]$ is linear and stable, it is possible to draw conclusions of the causal structure of a system from the CCF. In other words, if $u[k]$ and $y[k]$ have an excitation with the same frequencies but are time delayed, the CCF will reflect this in form of its amplitude and the time delay.

CCF of a 1st-order system   To illustrate the calculation of a CCF, figure 3.7 shows the 1st-order system introduced as a benchmark in section 3.2 while having $u[k]$ selected as white noise and $y[k]$ being the output time series. As for the benchmarks the sample rate is set to $T_s = 0.1\,$s. For $\lambda > 0$ the CCF shows a maximum at $\lambda = 1$ followed by a decaying exponential function, while for $\lambda \leq 0$ the CCF has values close to zero. This leads to the conclusion that there is a causal dependency pointing from $u \to y$ and not from $y \to u$.

The algorithm using the maximum amplitude of the cross-correlation function and the corresponding λ as a causal measure is explained in the next section. It is based on the concept to check if the found causal dependency is statistically significant and calculates at the same time the value of the causal strength of the dependency.

Using the cross-correlation function as a causal measure found already entry in some algorithms while one of the first ones was presented by Horch [Hor00] in 2000. The main purpose of this algorithm is to test if the maximum amplitude of the CCF calculated on the analyzed data set differs significantly from the maximum amplitude of the CCF resulting from two independent random variables.

This approach has been further refined by Bauer [Bau05] in 2005. In that case a compound parameter containing the maximum amplitudes of the CCF for $\lambda > 0$ and $\lambda < 0$ is calculated and tested in terms of a 3σ-test against the probability distribution of the compound parameter calculated from two independent random variables. The idea of the 3σ-test is to check if the value of the compound parameter for the non-permuted signal passes three times the value of the standard deviation generated from the compound parameter from random permutations of the input signal. In addition, in that algorithm a second test is performed, which checks if the resulting time shift for the two series differs significantly from zero.

In the following an algorithm is proposed which uses permutations of the original time series instead of random variables. Compared to the generation of random signals using permutations of the original series has the advantage that the characteristics of the amplitudes are kept for analysis while possible causal dependencies pointing from one time series to the other are destroyed.

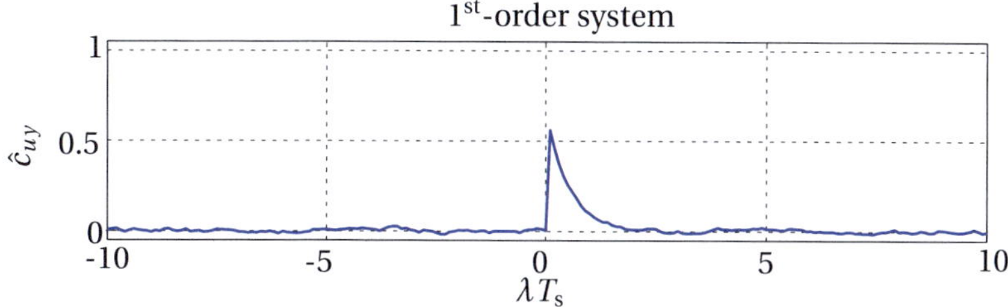

Figure (3.7) CCF of a 1st-order system. The decaying e-function for $\lambda > 0$ indicates a causal dependency from $u \to y$ while having the maximum amplitude at $\lambda T_s = 0.1\,\text{s}$.

3.4.1 Detection of Significant Causal Dependencies

When using the CCF for the detection of a cause-effect relationship $u \rightarrow y$ one of the main drawbacks is that it always detects a maximum. Therefore, it is important to establish two tests. The first test checks if the found absolute maximum or minimum value of the cross-correlation function differs significantly from zero or if this could also be the result from two uncorrelated time series. In the second test it is checked if the found causal strength from $u \rightarrow y$ differs significantly from $y \rightarrow u$ to have a clear indication of the causal flow. Furthermore, the value of the second test is used to describe the strength of the found cause-effect relationship.

1st test: Significant correlation of time delayed signals   To test if the absolute maximum amplitude of the CCF differs significantly from zero and is not generated through two uncorrelated signals, a hypothesis test which is based on the Pearson correlation coefficient [Joh93] is performed. The Pearson correlation coefficient is a traditional measure for the linear dependency of two different discrete time series and in that case is equal to the values of the cross-correlation function.

To perform the test, initially the maximum amplitude of the CCF needs to be selected. As it is possible that one of the signals is inverted, the absolute value of the CCF is used meaning that the position of the largest amplitude λ^{max} is defined as

$$\lambda^{\text{max}} := \arg\max_{\lambda} |\hat{c}_{uy}[\lambda]| \qquad \text{with } \lambda = 1 - K, \dots, -1, 1, \dots, K - 1. \tag{3.5}$$

If $\lambda^{\text{max}} > 0$ counts, this implies as there is a possible causal relation from $u \rightarrow y$ and the hypothesis test can be performed. Therefore, in a next step the value of the correlation coefficient $\hat{\rho}_{uy}$ needs to be selected, which is defined as

$$\hat{\rho}_{uy} := \hat{c}_{uy}[\lambda^{\text{max}}]. \tag{3.6}$$

This value will be used for the test. Due to the fact that the sample sizes of the time series are limited, equation 3.4 gives as a result only an estimate of the cross-correlation function. This means that $\hat{\rho}_{uy}$ will always differ slightly from zero when calculating it for two uncorrelated time series. Hence, the test is needed to check if a significant connection between the signals is present. According to [Hei10] the estimate of $\hat{\rho}_{uy}$ from two uncorrelated signals follows a Student's t-distribution with $N-2$ degrees of freedom, meaning that a classic t-test defined as

$$t_{\text{score}} = \hat{\rho}_{uy}\sqrt{\frac{N-2}{1-\hat{\rho}_{uy}^2}} \tag{3.7}$$

can be performed to check for a significant correlation. For this test, the null hypothesis H_0 is defined that there is no correlation between the two time series, meaning that if $t_{\text{score}} > t_{(N-2;1-\alpha/2)}$ counts, it can be assumed that the two signals are correlated.

The underscore of the t-distribution means that it has $N-2$ degrees of freedom and a probability of error of $1-\alpha/2$. In that case α is defined as threshold for statistical significance and throughout the thesis it is set to 0.05.

If the test indicates that at $\lambda^{\max}$ the null hypothesis can be rejected, which means that a significant correlation exists, a second test based on the maximum amplitudes for $\lambda < 0$ and $\lambda > 0$ is performed. If this test fails or $\lambda^{\max} < 0$ counts, it is assumed that no cause-effect relationship exists from $u \rightarrow y$ and the direction $y \rightarrow u$ can be tested.

2nd test: Significant causal direction   The aim of the first test was to check if the two signals are correlating with each other by selection the maximum value of the CCF. This test doesn't cover the possibility that the resulting CCF can have a global maximum for $\lambda > 0$ and a slightly lower local maximum for $\lambda < 0$ meaning that the underlying causal direction of the two signals is not obvious. When determining the outcome of the benchmark data sets (see section 3.4.2), this is for example the case when having data with a very narrow bandwidth or an instationarity in the data set.

Therefore, a second significance test is necessary which checks if the found maximum for $\lambda > 0$ is significantly different from the found maximum for $\lambda < 0$.

To perform the test, the compound parameter C^{CCF} defined as

$$C^{\mathrm{CCF}} := \frac{\max_{\lambda>0}|\hat{c}_{uy}[\lambda]| - \max_{\lambda<0}|\hat{c}_{uy}[\lambda]|}{\max_{\lambda>0}|\hat{c}_{uy}[\lambda]| + \max_{\lambda<0}|\hat{c}_{uy}[\lambda]|}, \tag{3.8}$$

with $-1 \le C^{\mathrm{CCF}} \le 1$ is calculated.

A value $C^{\mathrm{CCF}} > 0$ indicates a causal dependency from $u \to y$. As the resulting C^{CCF} strongly depends on the characteristics of $u[k]$ and $y[k]$, a dynamic threshold depending on $u[k]$ and $y[k]$ needs to be determined. This threshold is derived through a 3σ permutation test.

As a complete permutation of $u[k]$ destroys all causal information, the resulting value of C^{CCF} should be close to zero. This idea can be exploited by calculating random permutations of the input time series $u[k]$ and from this generate several values for the compound parameter, named C_{π}^{CCF}. In that case the index π indicates that permutations of the input time series have been used. Finally, the threshold $C_{\mathrm{thresh}}^{\mathrm{CCF}}$ for each data set is calculated as

$$C_{\mathrm{thresh}}^{\mathrm{CCF}} := \mu_{C_{\pi}^{\mathrm{CCF}}} + 3\sigma_{C_{\pi}^{\mathrm{CCF}}}, \tag{3.9}$$

with,

$$\mu_{C_{\pi}^{\mathrm{CCF}}} = \frac{1}{N}\sum_{n=1}^{N} C_{\pi}^{\mathrm{CCF}}[n], \qquad \sigma_{C_{\pi}^{\mathrm{CCF}}} = \sqrt{\frac{1}{N}\sum_{n=1}^{N}(C_{\pi}^{\mathrm{CCF}}[n] - \mu_{C_{\pi}^{\mathrm{CCF}}})^2}. \tag{3.10}$$

In that case $\mu_{C_{\pi}^{\mathrm{CCF}}}$ describes the empirical mean value and $\sigma_{C_{\pi}^{\mathrm{CCF}}}$ is the empirical standard deviation of C_{π}^{CCF}.

If the outcome indicates that $C^{\mathrm{CCF}} > C_{\mathrm{thresh}}^{\mathrm{CCF}}$ the found causal dependency is defined as being significant.

For the estimation of the pdf each time a total number of $N = 1000$ random permutations is used. Figure 3.8 shows the resulting pdfs using a kernel estimate for the 1st-order system in the benchmark data sets using once white noise and once white noise with superposed step function as input signal. The pdfs are generated for $K \in \{100, 1000, 10000\}$ samples. Comparing the two plots for white noise and the step function as input signals to each other, shows that the pdfs differ considerably. Furthermore, the impact of the sample size becomes obvious for the resulting pdf which shows again the need to generate a new threshold $C_{\text{thresh}}^{\text{CCF}}$ for each tested cause-effect relationship.

The resulting $C_{\text{thresh}}^{\text{CCF}}$ are given for white noise in order of the sample sizes as $0.35, 0.25, 0.18$ and for the step function as $0.46, 0.41, 0.37$. This again shows that the threshold needs to be adapted for each investigated pair $u[k], y[k]$.

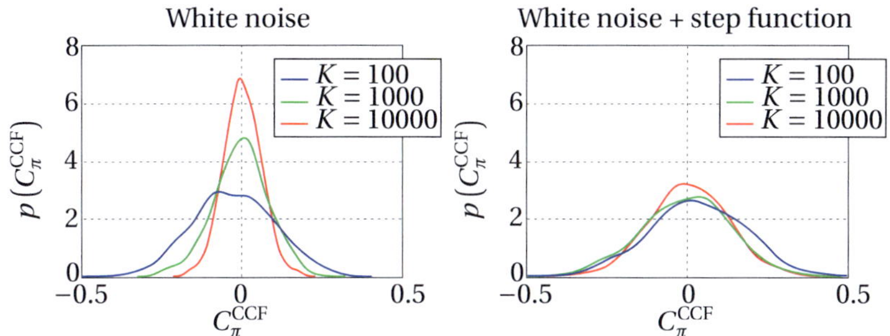

Figure (3.8) Resulting estimated pdfs of C_π^{CCF} for different signal lengths K. The left plot shows the pdfs for the 1st-order system using white noise, the right plot shows the pdfs for the step function as input signal.

Strength of causal dependency   The two defined tests check if there exists a significant causal dependency $u \to y$ but give no information about its strength. The causal strength of $u \to y$ can be constructed directly from C^{CCF}, defined as

$$Q^{\text{CCF}} := \max(0, C^{\text{CCF}})^{\beta_{\text{CCF}}}, \tag{3.11}$$

resulting in a value $0 \le Q^{\text{CCF}} \le 1$.

The parameter β_{CCF} is used as a tuning parameter and is set to one when testing the method on the benchmark data sets. As the selection of β_{CCF} depends on the behavior of other methods, a proper analysis on how to set β_{CCF} is given in chapter 4 after the other methods have been introduced.

As the CCF is a bivariate measure, for the generation of the causal matrix each variable is tested against any other. The proposed algorithm for the detection of cause-effect relationships for two time series $u[k]$ and $y[k]$ is summarized as algorithm 1.

Algorithm 1: Summary of the proposed algorithm based on the cross-correlation function

---

1. Compute $\hat{c}_{uy}$ of the two time series $u[k]$ and $y[k]$ and select $\lambda^{\max}$;

2. If $\lambda^{\max} < 0$ then $u \nrightarrow y$, else test if $t_{\text{score}} > t_{(N-2;1-\alpha/2)}$ counts;

3. Calculate C^{CCF} and the corresponding threshold $C^{\text{CCF}}_{\text{thresh}}$;

4. Check if $C^{\text{CCF}} > C^{\text{CCF}}_{\text{thresh}}$; set Q^{CCF} as resulting value of the causal strength $u \rightarrow y$

3.4.2 Tests with the Benchmarks

The proposed algorithm has been tested against the benchmarks introduced in section 3.2. For the calculation of the causal matrices the significance level α of the t-test is set to 0.05. The resulting cause-effect dependencies for each benchmark are given in figure 3.9 in terms of a doughnut chart where the expected causal dependency $u \rightarrow y$ is marked with a red square.

Figure 3.10 illustrates the generated cross-correlation function. This illustration is given to have further insights why causal dependencies could or could not be detected using the suggested approach.

The result of the **base configuration** shows that the causal dependency $u \rightarrow y$ is detected correctly. As explained in section 3.4 for a 1st-order system with white noise as input the resulting CCF consists of a decaying exponential function with its maximum amplitude at $\lambda = 1$. For negative λ the CCF consists of values close to zero. This leads to a large value of C^{CCF} and the causal dependency is correctly detected.

The cause-effect dependency using the **squaring device** cannot be found as the CCF is a linear measure. Hence, the plot of the CCF results over the whole range of λ in values close to zero.

The benchmark using the **feedback loop** is correctly detected while the maximum of the CCF is at $\lambda = 1$ and the causal strength leads to a slightly higher value than the one of the base configuration.

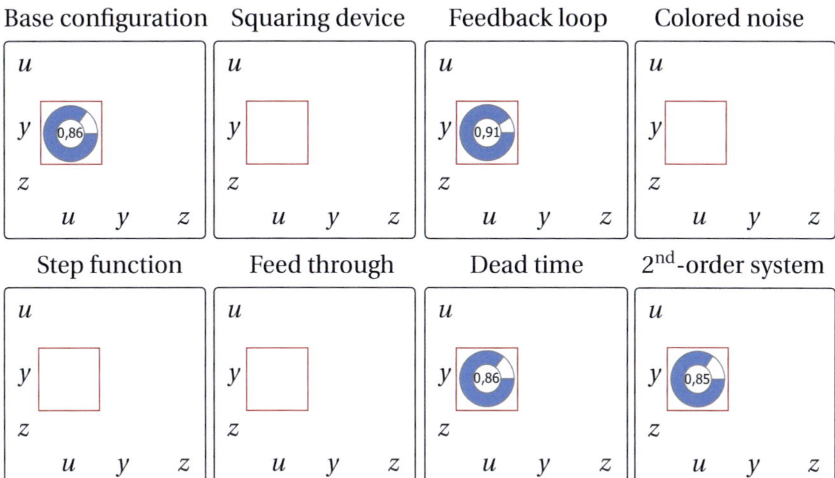

Figure (3.9) Visualization of the causal matrices in terms of doughnut charts for the developed algorithm tested on the benchmarks. The red square represents the expected causal dependency.

The CCF fails to detect the causal dependency of the benchmarks with **colored noise** and the **step function**. The reason is that in both cases the second significance test, which is based on the calculation of the significant causal direction, failed. This becomes obvious when regarding the plotted CCFs, as in both cases $\max_{\lambda<0}|\hat{c}_{uy}[\lambda]|$ and $\max_{\lambda>0}|\hat{c}_{uy}[\lambda]|$ are almost equal.

The **feed through** could not be detected as the cross-correlation has its maximum at $\lambda = 0$ and all other values of the function are around close to zero. In that case both significance tests fail.

The benchmark including a **dead time** doesn't affect the result of the CCF as this only leads to a difference at the position of $\lambda^{\max}$. Compared to the base configuration this means that the decaying e-function is delayed by 2 seconds which is the selected value of the dead time. The resulting causal strength is the same as for the base configuration.

Regarding the 2nd**-order system** the plotted CCF shows a different shape compared to the benchmarks based on 1st-order systems, but there exists a distinct difference between $\max_{\lambda<0}|\hat{c}_{uy}[\lambda]|$ and $\max_{\lambda>0}|\hat{c}_{uy}[\lambda]|$ resulting in the correct causal dependency $u \rightarrow y$. The causal strength has a slightly lower value compared to the base configuration.

Finally, the causal matrices show that the additionally added time series $z[k]$ from another process was in all cases correctly detected as a noise variable, as the algorithm did not find any cause-effect dependency pointing to or from it.

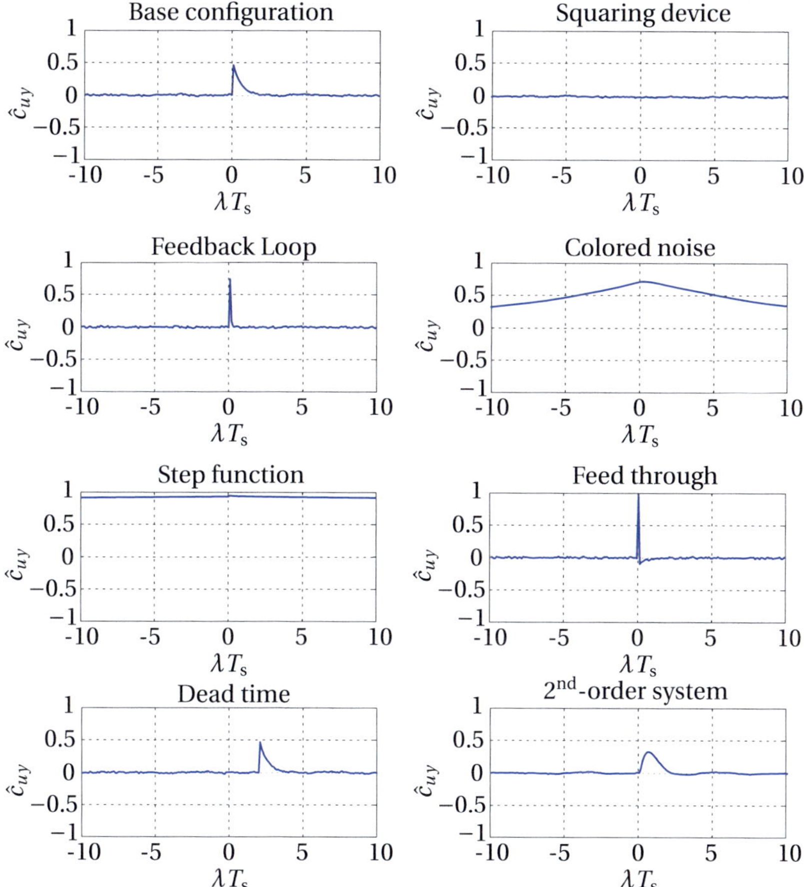

Figure (3.10) Visualization of the cross-correlation functions resulting from the benchmark data sets. This representation gives further insights in the used algorithm concerning its ability to detect causal dependencies in a data set.

3.5 Transfer Entropy

The concept of transfer entropy (TE) was first introduced by Schreiber [Sch00] in the year 2000 to describe the flow of information between two time series. From its definition it is a non-symmetric measure and can be used to detect causal dependencies in data by testing how much information is transferred from a time series $u[k]$ to a second series $y[k]$ and how much information is transferred back from $y[k]$ to $u[k]$. In difference to the CCF, which calculates correlations in data by time-shifting one of the series $u[k]$ or $y[k]$, the transfer entropy evaluates dependencies in terms of transition probabilities. This approach has the advantage that the TE can also be used to detect nonlinear cause-effect relationships in a data set. Applications of the TE are already found in various areas. There has been lots of work in the field of neuroscience [CMLVQ03],[SL09],[SGT$^+$09] but also in financial data analysis [MK02]. In the field of process engineering research has been made by Bauer [Bau05] who uses TE for the causal analysis on measurement data of chemical processes.

3.5.1 Entropic Measures for Static Probability Distributions

Entropic measures are used to calculate the information that is included in a data stream and were first introduced by Shannon [Sha48] in 1948 in statistical communication theory. Similar to its counterpart in thermodynamics, entropy in information theory can be understood as the amount of randomness or uncertainty in a data set. The main advantage when using these measures is that a signal is treated as the result of a random process where only the probabilities are evaluated. There is lots of recent literature and good introductions about entropic measures can be found e. g. in [CT06], [Pyl99] or [Mik09] which can give further insights regarding its application for data analysis. In the following the main concept is stated and it is explained why these entropic measures for static probability distributions cannot be used for causal analysis.

Entropy   For the calculation of entropy it is assumed that the time series $u[k] \in \mathbb{R}$ with $k = 1,\dots,K$ has been quantized having D_u discretization steps and values $i \in \{1,\ldots,D_u\}$. According to Shannon's definition the amount of information encoded in $u[k]$ is then written as

$$H_u = -\sum_{i=1}^{D_u} P(u = i) \log P(u = i), \tag{3.12}$$

which leads to values between $0 \leq H_u \leq \log D_u$. For the calculation of the logarithm usually the base 2 is used. If the base is set to 2, the unit in which the entropy is measured is then called bits. Regarding $u[k]$ there are two possible time series being on the boundaries.

An entropy of $H_u = 0$ is achieved by having a constant signal as in that case no information is encoded in the signal. In equation 3.12 this implies $P(u = i) \log P(u = i)$ which results in $1 \log(1) = 0$.

On the contrary, the signal with the highest entropy is uniform random noise. In that case every outcome is equally likely to occur. This can be explained by setting $P(u = i) = \frac{1}{D_u}$ into equation 3.12 which gives $H_u = \log D_u$ meaning that the time series has the highest uncertainty.

Joint entropy The joint entropy is an entropic measure which calculates the uncertainty contained in a set of variables. Taking into account a second time series $y[k]$ with quantified values $j \in \{1,\ldots,D_y\}$ and D_y describing the number of discretization steps, the joint entropy is given as

$$H_{u,y} = -\sum_{i=1}^{D_u}\sum_{j=1}^{D_y} P(u = i, y = j) \log P(u = i, y = j). \tag{3.13}$$

with the properties of the joint entropy given as $H_{u,y} \geq \max[H_u, H_y]$ and $H_{u,y} \leq H_u + H_y$.

Conditional entropy Conditional entropy is defined as the uncertainty left in a random variable when the outcome of another random variable is known. Calculating the uncertainty of $y[k]$, when knowing the outcome of $u[k]$, is given as

$$H_{y|u} = \sum_{i=1}^{D_u}\sum_{j=1}^{D_y} P(u = i, y = j) \log \frac{P(u = i)}{P(u = i, y = j)}, \tag{3.14}$$

after quantification of the two series. The conditional entropy takes values in the interval $0 \leq H_{y|u} \leq H_y$ and $H_{u|y}$ is defined analogous. $H_{y|u} = 0$ means that the information in $y[k]$ is completely incorporated in $u[k]$, while $H_{y|u} = H_y$ means that $u[k]$ has no information about $y[k]$.

Conditional entropy is an asymmetric measure, meaning that $H_{y|u} \neq H_{u|y}$. This could lead to the delusive conclusion that it can be used as a causal measure. Still, as e.g. shown in [CT06], the equations 3.12, 3.13 and 3.14 are linked together as

$$H_{y|u} = H_{u,y} - H_u,$$

(3.15)

$$H_{u|y} = H_{u,y} - H_y.$$

(3.16)

These equations can then again be transformed into

$$H_u - H_y = H_{u|y} - H_{y|u}.$$

(3.17)

The resulting equation 3.17 reveals that the difference of the two conditional entropies is the same as the difference of two entropies. In other words, no causal information can be drawn from the calculation of conditional entropies even if it is an asymmetric measure.

For this reason the concept of transfer entropy is needed which is based on the calculation of transition probabilities. The concept and its application for the detection of cause-effect relationships is explained in the following section.

3.5.2 Transfer Entropy as a Causal Measure

As stated, entropic measures based on static probability distributions cannot be used to reconstruct the propagation path of a plant-wide disturbance. The reason is that these measures do not take dynamical information of the ana-lyzed time series into account. In contrary to these traditional measures, the transfer entropy uses transition probabilities and thus dynamical information, to figure out to which extend two signals exchange information. For $y[k]$, the transition probability is defined as $P(y_{n+1}|\boldsymbol{y})$. This is used as short notation of

$$P(y_{n+1} = j_{n+1}|y_n = j_n,...,y_1 = j_1), \tag{3.18}$$

with n defining the time horizon (which is equivalent to the model order), used for the calculation of the transition probability. Since the transfer entropy deals with the calculation of probabilities it has to be noticed that y_{n+1} should not be confounded with $y[k+1]$. The variable j describes again the quantization levels of $y[k]$. In addition to $P(y_{n+1}|\boldsymbol{y})$, for the calculation of the TE, the transition probability $P(y_{n+1}|\boldsymbol{y},\boldsymbol{u})$ needs to be calculated. This is again the short notation of

$$P(y_{n+1} = j_{n+1}|y_n = j_n,...,y_1 = j_1,u_n = i_n,...,y_1 = i_1). \tag{3.19}$$

The transfer entropy describing the information flow from the time series $u[k]$ to $y[k]$ is then defined as the information y and u contain about future values of y, subtracted by the information y has about itself for prediction. Writing this in a probabilistic manner, the transfer entropy is then given as

$$\mathrm{TE}^\star_{uy}(\lambda) := \sum_{\substack{j_1=1 \\ \cdots \\ j_{n+1}=1}}^{D_y} \sum_{\substack{i_1=1 \\ \cdots \\ i_n=1}}^{D_u} P(y_{n+1},\boldsymbol{y},\boldsymbol{u}) \log \frac{P(y_{n+1}|\boldsymbol{y},\boldsymbol{u})}{P(y_{n+1}|\boldsymbol{y})}. \tag{3.20}$$

According to Marschinski [MK02], the range of the transfer entropy is the interval $0 \leq \mathrm{TE}^\star_{uy} \leq H_y$.

To capture dead times in the data, the parameter λ is introduced to perform a backward-shifting of $u[k]$. Hence, equation 3.20 is calculated for different $u[k-\lambda]$. Capturing dead times through an increase of the time horizon is not appropriate as this can lead to numerical problems for large values of n. In the following as a short-hand definition $\mathrm{TE}_{uy} := \max_\lambda(\mathrm{TE}^\star_{uy}(\lambda))$ is used, which means that the maximum value of the transfer entropy as a function of λ is taken.

Effective transfer entropy The effective transfer entropy (ETE) has been introduced by Marschinski [MK02] and deals with the problem of artifacts resulting from the calculation of the TE. In detail, these originate from the finite sample size of the investigated time series and the estimation of the probability distributions. Both effects have a negative impact on the resulting transfer entropy, as even for two independent series the TE results in values larger zero. Compared to the calculation of the transfer entropy the ETE takes this into account and is given as

$$\mathrm{ETE}^\star_{uy}(\lambda) = \mathrm{TE}^\star_{uy}(\lambda) - \mathrm{TE}^\star_{u_\pi y}(\lambda). \tag{3.21}$$

In this equation $\mathrm{TE}^\star_{u_\pi y}(\lambda)$ is the TE calculated from a permutation of the time series $u[k]$. Like for the CCF, this means that all causal dependencies $u \rightarrow y$ are broken and resulting values different from zero in $\mathrm{TE}^\star_{u_\pi y}(\lambda)$ are the outcome of numerical issues. Similar to the definition of a threshold for the CCF, this approach is used to calculate a threshold based on $\mathrm{TE}^\star_{u_\pi y}(\lambda)$ to detect only significant causal dependencies in a data set. This is explained in the next section.

3.5.3 Detection of Significant Causal Dependencies

Using transfer entropy for the detection of causal dependencies requires several user-selected parameters. The data quantization method of the signals used for analysis has to be selected and the parameter n for the size of the time horizon needs to be set. Furthermore, a significance test has to be defined and finally the strength of the found cause-effect relationship needs to be calculated. Performing an appropriate selection of these parameters is explained in this section.

Data quantization   As stated, in the previous sections as a first step for the calculation of the transfer entropy the time series $u[k]$, $y[k]$ need to be quantized to calculate the probability distributions. In the proposed algorithm this is done by using an approach suggested by [MK02]. In so doing, the range of values of the time series $u[k]$ is divided into D_u intervals, with the condition, that in each interval the number of data points is equal. As a next step the value of every data point is replaced by the label of its interval.

In terms of an entropic measure this means that $P(u = i) = \frac{1}{D_u}$ counts which leads to the entropy $H_u = -D_u \frac{1}{D_u} \log \frac{1}{D_u} = \log D_u$ of the quantized time series. In other words quantization of the data in this way leads to maximum entropy. The output series $y[k]$ is quantized analogously. One further advantage using this approach is that effects coming from skewed data are corrected as the resulting histogram of the series is uniform.

Selection of the time horizon To select the time horizon (or the model order) used for calculating the transition probabilities, the parameter n needs to be set accordingly. In order to do this, two requirements need to be taken into account. Setting the parameter n too small means that possibly not the complete system dynamics are covered by the time horizon which can lead to the detection of false causal dependencies.

Selecting n too large can lead to problems as the calculation of the transition probabilities becomes computationally more expensive as more combinations in $P(y_{n+1}|\boldsymbol{y})$ and $P(y_{n+1}|\boldsymbol{y},\boldsymbol{u})$ need to be evaluated. Additionally, it has to be taken into account when having only short time series, a time horizon $n > 1$ means a sub-sampling of the data which reduces the data set size further. Hence, the value of n should be as large as needed to cover all the system dynamics and as small as possible to reduce the computational burden.

As the system dynamics are captured through the output series $y[k]$, the model order n is estimated by analyzing this series. Therefore, the residual sum of squares (RSS) of several vector autoregressive (VAR) models are calculated, where the RSS vor a specific model order n results in

$$\hat{\sigma}_y^2 = \sum_{k=n+1}^{K} \left(y[k] - \hat{a}_0 - \sum_{j=1}^{n} \hat{a}_j y[k-j] \right)^2. \tag{3.22}$$

The values of the parameters $\hat{a}_j$ result from a least square estimation. Each VAR model is calculated using $n = 1,\ldots,n_{\max}$ as model order with $n_{\max}$ being set as the maximum model order. The finally used order n_{TE} for the transfer entropy is chosen as the minimum of the Akaike information criterion (AIC) [Aka74]. This is calculated as

$$\mathrm{AIC}(n) = \log \hat{\sigma}_y^2 + \frac{2n}{K}. \tag{3.23}$$

For illustration, figure 3.11 shows the resulting TE_{uy} for the 1st-order system with squaring device and the 2nd-order system for $K \in \{200, 500, 1000\}$ samples depending on the model order n. The outcome shows that smaller sample sizes also lead to lower values for the transfer entropy. When increasing the time horizon n this leads to a similar result. The reason is, that more data is needed to estimate the transition probabilities $P(y_{n+1}|\boldsymbol{y},\boldsymbol{u})$ and $P(y_{n+1}|\boldsymbol{y})$ in equation 3.20 if the time horizon is set on a large value. As expected from system dynamics, the 1st-order system has its maximum at $n = 1$, the 2nd-order system at $n = 2$. Using the AIC criterion, the model order n_{TE} for the 1st-order system with squaring device results for all sample sizes as $n_{\text{TE}} = 2$ and for the 2nd-order system as $n_{\text{TE}} = 4$. This reveals that the AIC slightly overestimates the correct model order as it doesn't find the maximum position of the TE but still selects model orders resulting in large values of the transfer entropy.

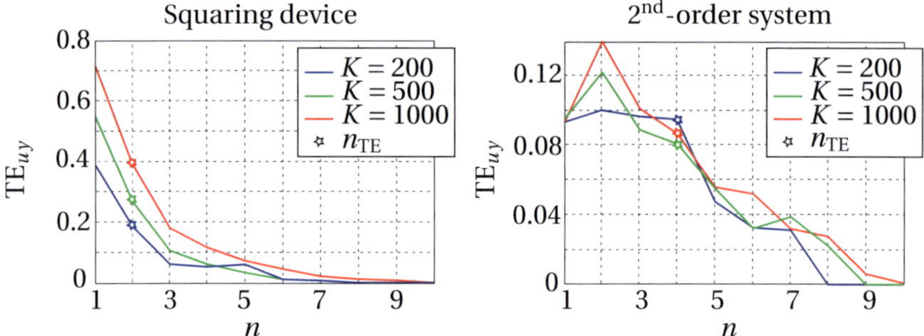

Figure (3.11) TE_{uy} for different sample sizes K plotted over the model order n. An increase in n or a small sample size results in low values of TE_{uy}. The selected order n_{TE} through the AIC criterion is marked with a cross in the plots.

Calculating the significance threshold   Having a low value of the transfer entropy indicates that no causal dependency exists between two time series, while large values indicate a strong relationship. To put this into a statistical measure, when testing for a significant cause-effect dependency, a significance test similar to the test used for the cross-correlation function in section 3.4 is performed. Regarding the transfer entropy this test has been originally suggested by Schreiber [Sch00] and Bauer [Bau05]. The key idea is to generate a threshold $\mathrm{TE}_{uy}^{\mathrm{thresh}}$ based on the permutated input time series $u_\pi[k]$. As explained when introducing the ETE a permutation of $u[k]$ destroys all causal dependencies $u \rightarrow y$ and hence $\mathrm{TE}_{u_\pi y}$ should result in 0 but due to numerical artifacts this is not the case. This characteristic is exploited by permuting $u[k]$ N times while each time $\mathrm{TE}_{u_\pi y}$ is calculated. The resulting values of $\mathrm{TE}_{u_\pi y}$ are used as an estimate of a probability density function and to calculate the threshold $\mathrm{TE}_{uy}^{\mathrm{thresh}}$ in terms of a 3σ-test. This is defined as

$$\mathrm{TE}_{uy}^{\mathrm{thresh}} := \mu_{\mathrm{TE}_{u_\pi y}} + 3\sigma_{\mathrm{TE}_{u_\pi y}}. \tag{3.24}$$

If $\mathrm{TE}_{uy} > \mathrm{TE}_{uy}^{\mathrm{thresh}}$ holds, the significance test has passed and it is assumed that a cause-effect relationship exists and pointing from $u \rightarrow y$. For the estimation of the probability distribution each time $N = 1000$ random permutations are used. Exemplarily figure 3.12 illustrates the estimated pdfs for $\mathrm{TE}_{u_\pi y}$ for different samples sizes K when having white noise and colored noise as input signal.

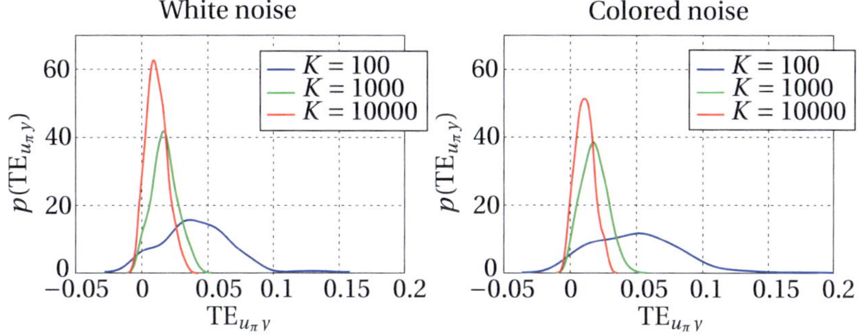

Figure (3.12) Resulting pdfs of $\mathrm{TE}_{u_\pi y}$ for different signal lengths K. The left plot shows the pdfs for the 1st-order system using white noise, the right plot shows the pdfs when having colored noise as input signal.

Calculation of causal strength Besides having the information if a causal dependency is significant, the causal strength needs to be calculated. As explained in section 3.5.2 the transfer entropy has values in $0 \leq \mathrm{TE}_{uy} \leq H_y$, thus the causal strength of the transfer entropy can be normalized as

$$Q^{\text{TE}} := \left(\frac{\text{TE}_{uy}}{H_y} \right)^{\beta_{\text{TE}}} , \tag{3.25}$$

resulting in a value between $0 \leq Q^{\text{TE}} \leq 1$.

As for the cross-correlation function the parameter β_{TE} is a tuning parameter which needs to be set when combining the different methods to one causal matrix. Regarding the benchmarks presented in the next section, β_{TE} is set to one. A proper analysis on how to fit β_{TE} will be given in chapter 4 after all methods have been introduced.

As the transfer entropy is a bivariate measure each process variable needs to be tested against each other for the generation of the causal matrix. Finally, the suggested algorithm for the detection of cause-effect relationships is given in algorithm 2.

Algorithm 2: Summary of the proposed algorithm based on transfer entropy

---

1. Compute model order n_{TE} of the transfer entropy using VAR models and AIC;

2. Compute $\text{TE}^{\star}_{uy}(\lambda)$ of the time series $u[k]$ and $y[k]$ and set $\text{TE}_{uy} = \max(\text{TE}^{\star}_{uy}(\lambda))$;

3. Calculate $\text{TE}^{\text{thresh}}_{uy}$;

4. Check if $\text{TE}_{uy} > \text{TE}^{\text{thresh}}_{uy}$; set Q^{TE} as the resulting value of the causal strength $u \rightarrow y$

3.5.4 Tests with the Benchmarks

In this section the proposed approach to use transfer entropy for fault localization and the calculation of the disturbance propagation path is tested on the benchmarks introduced in section 3.2. The resulting causal dependencies for each benchmark are presented in terms of a doughnut chart in figure 3.13. In addition, like for the CCF, figure 3.14 illustrates the transfer entropy over several

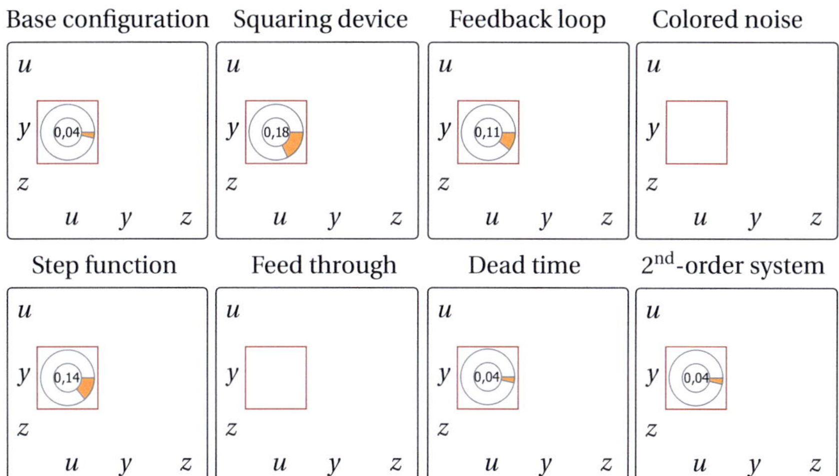

Figure (3.13) Causal matrices from the developed algorithm for the detection of the causal dependency for the tested benchmarks. The red squares indicate the expected causal dependency.

values of λT_s. Showing the TE over λT_s has the aim to give further insights why some cause-effect relationships could or could not be detected using the proposed algorithm. Therefore, the transfer entropy for $\lambda < 0$ indicates the causal dependency from $y \to u$, while for $\lambda > 0$ the causal dependency pointing from $u \to y$ is shown.

Reviewing the output of the benchmarks indicates that the causal dependency using the **base configuration** has been correctly detected where $\max(\mathrm{TE}^\star_{uy}(\lambda))$ has its position at $\lambda = 1$. Calculating the causal dependency from $y \to u$ yields values close to zero with a maximum value of the transfer entropy below the threshold.

The **squaring device** has no negative impact regarding the detection of the cause-effect relationships. This follows from the definition of the TE as it detects causal dependencies by calculating transition probabilities. Hence, it does not rely on linear dependencies and the correct causal relationship is found for this benchmark. Furthermore, the causal strength is larger compared to the base configuration due to a higher magnitude of $\max(\mathrm{TE}^\star_{uy}(\lambda))$.

The benchmark, using the data from the **feedback loop** shows the same results as the base configuration with a maximum value of the transfer entropy at $\lambda = 1$ and values close to zero for $\lambda < 0$. Therefore, the implied causal dependencies are correctly found as $u \rightarrow y$ and $y \nrightarrow u$.

The TE failed to detect a causal dependency $u \rightarrow y$ when testing it with the **colored noise**. Compared to the other benchmarks the maximum transfer entropy for the cause-effect relationship $u \rightarrow y$ is much lower resulting in $\mathrm{TE}_{uy} < \mathrm{TE}_{uy}^{\mathrm{thresh}}$. Additionally, it has to be mentioned that the colored noise has no impact on the transfer entropy for $\lambda < 0$ which means that no false causal direction $y \rightarrow u$ is found.

The data set containing white noise with a superposed **step function** is correctly found as the causal dependency $u \rightarrow y$ has been detected with a significant peak at $\lambda = 1$.

Regarding the **feed through** the algorithm could not find the cause-effect relationship $u \rightarrow y$ since the values of the TE stayed close to zero for all λ.

A **dead time** does not have an effect on the resulting transfer entropy as only $\max(\mathrm{TE}_{uy}^{\star}(\lambda))$ is shifted by $2\,\mathrm{s}$ compared to the base configuration but stays on the same value.

Regarding the data set from the $2^{\mathbf{nd}}$**-order system**, the TE plotted over λT_{s} shows a different shape compared to the base configuration as it has a maximum value at $\lambda = 2$. Hence, the causal dependency $u \rightarrow y$ is correctly found.

Finally, the resulting causal matrices show that the additionally added time series $z[k]$ coming from another process was in all cases correctly detected as noise variable. In all benchmarks the proposed algorithm did not indicate any cause-effect relationships pointing towards or from $z[k]$ meaning that in this case the resulting transfer entropy was always below the calculated threshold.

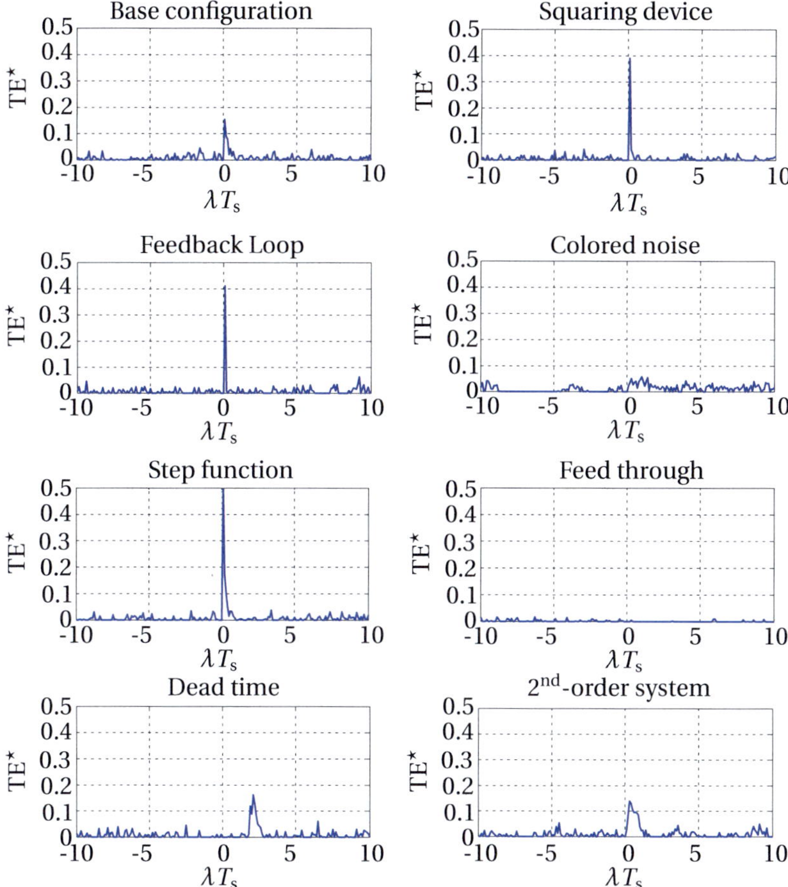

Figure (3.14) Visualization of the transfer entropy using the benchmark data sets. This representation gives further insights in the proposed algorithm regarding its ability to detect causal dependencies in a data set and in order to use it for fault localization.

3.6 Granger Causality

The concept of Granger causality (GC) has been originally introduced in the field of economics by Clive Granger in 1969 [Gra69] who used it to determine the relationships of different econometric models. Granger causality is based on a multivariate causality test and therefore the time series $u[k]$ is extended for this causality measure to a set of input variables which consists of $u_i[k], i = 1, \ldots, r$ time series. The basic concept of GC can be explained by assuming that initially only the two time series $u_i[k] \in \mathbb{R}$ and $y[k] \in \mathbb{R}$ with K samples are available. In that case, the causal influence $u_i \to y$ is assumed to exist if past values from $u_i[k]$ and $y[k]$ result in a higher accuracy in forecasting $y[k]$ than using only past values from $y[k]$. Figure 3.15 sketches the approach. This is evaluated mathematically by generating and comparing two linear vector autoregressive models, while the one containing only $y[k]$ is called the restricted and the one containing $u_i[k], y[k]$ is called the unrestricted model.

Traditionally, GC is used in the field of economics e. g. when analyzing consumer and business confidence [GC10] or when analyzing the relation between dividend yields and interest rates [Sau79]. Still, in recent time the application of GC is of growing interest especially in the field of neuroscience, where it is used to extract the directed functional connectivity from neural signals. An overview about how GC is used in this field is given in [DCB06]. Further interesting applications are the calculation of the causal influence of two neural system on each another [XKR$^+$10] and the calculation of the causal flow of protein-signaling networks [YL10]. In process technology, Yang [YX12] proposes in a review article to use Granger causality among other methods for a causal analysis of large-scale processes.

One-step-ahead prediction As previously mentioned, Granger causality is a multivariate measure. This means, when testing for a causal dependency $u_i \to y$ the series $y[k]$ can also causally depend on other time series $u_l[k]; l \neq i$ in the data set. GC takes this into account when designing the restricted and the unrestricted model.

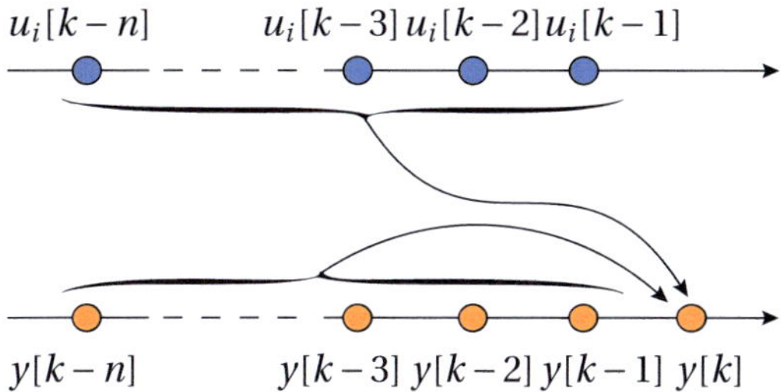

Figure (3.15) Illustration of the basic concept of Granger causality based on model comparison. If a time series $u_i[k]$ increases the prediction accuracy of $y[k]$ compared to using only past values of $y[k]$, it is said that $u_i[k]$ has a causal impact on $y[k]$.

A comparison of the models is done in terms of a one-step-ahead prediction. The concept is to predict the value of $y[k]$ for the next sample point exclusively by knowing previous observations. If the squared sum of the prediction error using the unrestricted model, including all time series $\{u_1[k],\ldots,u_r[k]\}$, is substantially smaller than the prediction error of the restricted model, taking into account only the time series $\{u_1[k],\ldots,u_r[k]\}\setminus u_i[k]$, it can be concluded that the unrestricted model has a significant better forecasting performance. This means that there exists a causal dependency $u_i \to y$.

For the proposed algorithm each time series is once selected as being the output $y := u_m$, while the left $r-1$ time series are used as input. The calculation of the residual sum of squares can be formulated with n defining the order of the vector autoregressive model and K the sample size as

$$E_{\overline{U}_i Y} = \sum_{k=n+1}^{K} \left(y[k] - \hat{a}_0 - \sum_{j=1}^{n} \hat{a}_j y[k-j] - \sum_{\substack{l=1,\\ l\neq i,m}}^{r} \sum_{j=1}^{n} \hat{b}_{lj} u_l[k-j] \right)^2, \qquad (3.26)$$

$$E_{U_i Y} = \sum_{k=n+1}^{K} \left(y[k] - \hat{a}_0 - \sum_{j=1}^{n} \hat{a}_j y[k-j] - \sum_{\substack{l=1,\\ l\neq m}}^{r} \sum_{j=1}^{n} \hat{b}_{lj} u_l[k-j] \right)^2, \qquad (3.27)$$

with

$$y[k] := u_m[k]. \qquad (3.28)$$

By definition, the parameters $\hat{a}_j$, $\hat{b}_{lj}$ in (3.26) and (3.27) result from separate estimations. As shown, the impact of the i-th input signal on the m-th output signal is measured in terms of the sum of the squares of residuals without u_i as $E_{\overline{U}_i Y}$ and with u_i as $E_{U_i Y}$. The values of $E_{\overline{U}_i Y}$ and $E_{U_i Y}$ are used to test for causal significance and calculating the causal strength. The selection of the correct model order n and the validation of the model consistency are described below.

3.6.1 Model Order Estimation and Consistency

As already shown for the transfer entropy, the performance of a model strongly depends on the selected model order. Regarding Granger causality, when estimating an autoregressive model, the time horizon n represents the model order which needs to be chosen appropriately. Selecting a too small n leads to a poor model with a large prediction error, setting n too large results in an overfitted model which works well on the trained data but not on a new data set.

To solve this problem, commonly two approaches are used to define n, namely the Akaike information criterion (AIC) [Set10] and the Bayesian information criterion (BIC) [Ris78]. The AIC has already been introduced in its basic form in section 3.5.3 for the estimation of the time horizon of the transfer entropy. In that case only the output time series was used for the selection of the model order. This is in difference to Granger causality as in that case the autoregressive model can be used directly for evaluation. AIC and BIC penalize the model complexity while taking into account the resulting prediction errors $E_{U_i Y}$ or $E_{\overline{U}_i Y}$, the sample size K, the number of variables r and the selected model order n. For model order estimation the loss function V is calculated by using the residual sum of squares of one of the two autoregressive models. Hence, the loss function is set to $V = E_{U_i Y}$ for the unrestricted and to $V = E_{\overline{U}_i Y}$ for the restricted model. Finally, the information criteria AIC and BIC for one autoregressive model is calculated as

$$\text{AIC(n)} = \log(V) + \frac{2np^2}{K}, \tag{3.29}$$

$$\text{BIC(n)} = \log(V) + \frac{\log(N)np^2}{K}. \tag{3.30}$$

There has been lots of research about comparing these two criteria for model averaging. For an overview it is referred to Burnham and Anderson [BA02]. Used in practice, both criteria show similar results. In this work AIC is used when selecting the order n while the maximum model order is set to $n_{\max} = 20$. The resulting model order from the AIC criterion for Granger causality is named n_{GC}.

Figure 3.16 shows the impact when varying n on the residual sum of squares of the prediction error for the 1st-order system with colored noise as input signal and for the 2nd-order system. For simulation $K = 1000$ samples are used. The outcome shows that in both cases it is possible to detect the correct underlying cause-effect relationship using GC (see section 3.6.3 for a thorough explanation). Selecting a larger n reduces in both cases the residual sum of squares for the restricted and for the unrestricted model. For the 1st-order system with colored noise as input the AIC believes a model order of 2, for the second-order system AIC leads to a model order of 12. In the first case the selected model order has no impact because the offset of the prediction error between the restricted and the unrestricted error stays almost constant. In case of the second-order model a too small model order would have resulted in an insignificant difference in the prediction error meaning that no causal dependency would have been detected.

Durbin-Watson statistic In a next step all estimated models need to be validated if they fit the underlying data set sufficiently. Therefore, a Durbin-Watson statistic [DW50] is applied which tests if the residuals, named as $\epsilon[k]$, of the estimated model are all uncorrelated. This should be the case for a correctly

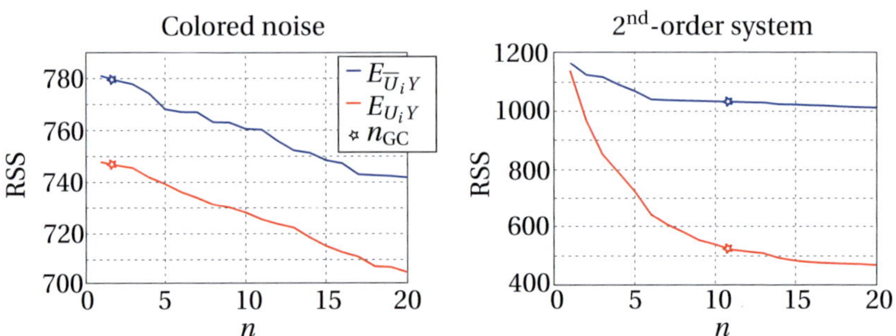

Figure (3.16) Impact of the selected model order on the residual sum of squares (RSS) when using Granger causality as causal measure. The selected model order n_{GC} through the AIC criterion is marked with a cross in the plots.

estimated model. The test statistics are calculated as

$$d = \frac{\sum_{k=2}^{K}(\epsilon[k] - \epsilon[k-1])^2}{\sum_{k=2}^{K}(\epsilon[k])^2}, \tag{3.31}$$

with $d \in [0,4]$. The limits of the measure describe for $d = 0$ a perfect positive autocorrelation and for $d = 4$ perfect negative autocorrelation. As a rule of thumb Seth [Set10] suggests that $d \stackrel{!}{\in} [2,3]$ should be fulfilled to consider the model as consistent. For each estimated model a Durbin-Watson statistic is performed and models that do not fulfill this criterion are not considered for causal analysis and the causal dependency is set to non-existing.

3.6.2 Significance and Causal Strength

A significance test needs to be performed by testing if $E_{\overline{U}_i Y}$ and $E_{U_i Y}$ differ significantly. Following [SH06] and [Set10], under the assumption that $E_{\overline{U}_i Y}$ and $E_{U_i Y}$ follow χ^2 distributions, an F-test can be performed to verify if the time series $u_i[k]$ has a causal influence on $y[k]$. The test is performed on the restricted and unrestricted model under the hypothesis that $E_{U_i Y} < E_{\overline{U}_i Y}$. For this case, the F-test is calculated as

$$F(n, K - n - p) = \frac{E_{\overline{U}_i Y} - E_{U_i Y}}{n} \cdot \frac{K - n - pn}{E_{U_i Y}}. \tag{3.32}$$

When performing the F-test, the significance level α needs to be defined which is set throughout the thesis to $\alpha = 0.05$.

Causal strength   The causal strength Q^{GC} is defined through a comparison of the two different squared sums of residuals $E_{\overline{U}_i Y}$ and $E_{U_i Y}$. Again a tuning parameter β_{GC} is used and regarding the benchmark data sets, it is set to one. The setting of the parameter is postponed to section 4.2. The resulting value Q^{GC} is defined as

$$Q^{GC} := \left(1 - \frac{E_{U_i Y}}{E_{\overline{U}_i Y}}\right)^{\beta_{GC}}, \tag{3.33}$$

with $0 \leq Q^{GC} \leq 1$, where 0 indicates no causal dependency and 1 stands for a reliably detected causal relationship. The suggested algorithm for using Granger causality to detect cause-effect relationships is summarized as algorithm number 3 .

Algorithm 3: Summary of the proposed algorithm based on Granger causality

---

1. Compute $E_{\overline{U}_i Y}$ and $E_{U_i Y}$ using the model order estimated through AIC or BIC;

2. Test for model consistency for both models using Durbin-Watson statistic;

3. Perform a significance test based on an F-test for $E_{U_i Y} < E_{\overline{U}_i Y}$;

4. If causal dependency is significant, set Q^{GC} as the resulting value of the causal strength $u_i \rightarrow y$;

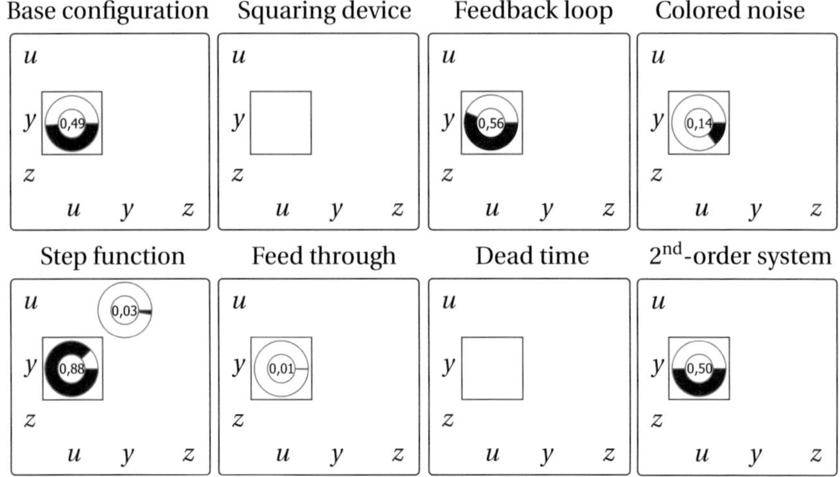

Figure (3.17) Causal matrices represented as doughnut charts when using Granger causality for the detection of the causal direction of the tested benchmark data sets. The red squares indicate the known causal dependency in the data set.

3.6.3 Tests with the Benchmarks

In the following, the behavior of Granger causality is investigated on the benchmark data sets presented in chapter 3.2. The resulting cause-effect dependencies are outlined in figure 3.17 in terms of a doughnut chart for each benchmark. Reviewing the performance, starting with the **base configuration** the results show that the causal dependency $u \rightarrow y$ has been correctly detected. All significance tests are passed and Q^{GC} yields a value of 0.49.

The nonlinear benchmark containing the **squaring device** cannot be detected when applying the algorithm. The reason is that the used vector autoregressive models from equations 3.26 and 3.27 assume a linear process structure.

The **feedback loop** has no negative impact on the found causal dependencies of the benchmark using this method. Furthermore, the causal strength results in a slightly larger value than the base configuration.

In contrast to the cross-correlation function and the transfer entropy, when having the benchmark with the **colored noise**, using GC it is possible to detect the causal dependency $u \rightarrow y$. In that case the causal strength $u \rightarrow y$ is lower than for the base configuration but still significant.

The causal dependency $u \rightarrow y$ in the benchmark containing white noise with a superposed **step function** can also be detected. Comparing this result to the one of the base configuration yields a much higher causal strength. Furthermore, a weak false causal relationship $y \rightarrow u$ has been found. This shows the advantage of using the causal strength as a quantitative measure since the value is close to zero. Detecting causal dependencies only based on significance tests would result in an equal intercausal relationship meaning that the two variables would have the same causal influence onto each other.

For the benchmark containing the **feed through**, GC detects a significant cause-effect relationship with a low causal strength. Compared to the other proposed algorithms, GC is the only method which detects a significant causal dependency $u \rightarrow y$ for this benchmark.

The cause-effect relationship in the benchmark containing a large **dead time** could not be found. This follows from the definition of GC in equations 3.26 and 3.27. Granger causality is not suited for the detection of causal dependencies containing large dead times as it can compensate them only through an increase of the model order n. As n_{max} is set to 10 but at least a model of order $n = 21$ ($T_{\mathrm{d}} = 2\,\mathrm{s}$, $T_{\mathrm{s}} = 0.1\,\mathrm{s}$) is needed, the causal dependency is not found. Setting the value of n_{max} too large is inappropriate since this can lead to computational problems when estimating the vector autoregressive models, especially when there is a large number of process variables.

Regarding the 2nd-**order system** the causal dependency $u \rightarrow y$ is found correctly and Q^{GC} devotes nearly the same causal strength as in the base configuration.

The causal matrices for all benchmarks show that the additionally added time series $z[k]$ coming from another process was correctly detected as noise variable. In all benchmarks the Granger causality did not indicate any cause-effect relationships pointing towards or from $z[k]$ meaning that in these cases the hypothesis tests failed.

3.7 Support Vector Machines

Support vector machines (SVM) are learning methods that are used for supervised learning and can be applied for classification and regression tasks. Originally they were developed by Vapnik [Vap82, Vap98] at the AT&T Bell Laboratories. In the beginning, the main interest was to use SVMs as classifiers for optical character recognition and object detection but later on where extended towards regression and time series prediction [SS04]. For both, classification and regression, SVMs proved to be competitive with other state-of-the-art machine learning methods which is outlined e. g. in [OD08, CST00, BFSS03], and were identified in 2007 [WKRQ$^+$07] as one of the 10 most influential data mining algorithms. Good tutorials about support vector machines used for classification can be found in [CST00, SS01], for regression tasks in [Vog08, SS04]. In recent times SVMs are also used in process technology where their main task is the condition monitoring of plants [TS09].

In this thesis a concept which uses SVMs for regression tasks in combination with variable selection and model reduction is proposed for the detection of cause-effect relationships and the reconstruction of the disturbance propagation path of a fault. Like for Granger causality, to test if a time series $u[k]$ causes $y[k]$, both series are used to predict future values of $y[k]$. Therefore, from both series the last n past values are selected to generate an input data set consisting of $\{u[k-1],...,u[k-n],y[k-1],...,y[k-n]\}$ while setting $y[k]$ as output for the SVM. In the next step, a variable elimination for the SVM is performed with regard to the prediction accuracy of $y[k]$. The variable elimination removes irrelevant or redundant data and finally results in a subset of relevant input variables for predicting $y[k]$. This increases the performance of the SVM and deals as a test of significance. If this subset still contains one or several past values of $u[k]$, it is assumed that u causes y. Like for the autoregressive models when testing for Granger causality a Durbin-Watson statistic is performed to validate if the resulting SVM fits the data sufficiently.

As the performance of the SVM depends on several user-selected parameters the advantage of this approach is that only one SVM is needed for the detection of a significant causal dependency (for the calculation of the causal strength two SVMs with the same parameter settings are used). Comparing two SVMs with different user-selected parameters, e. g. the first SVM only with past values of $y[k]$ as input and the second SVM with past values of $u[k]$ and $y[k]$ as input, can lead to misleading results due to possibly bad fitted parameters for one of the SVMs.

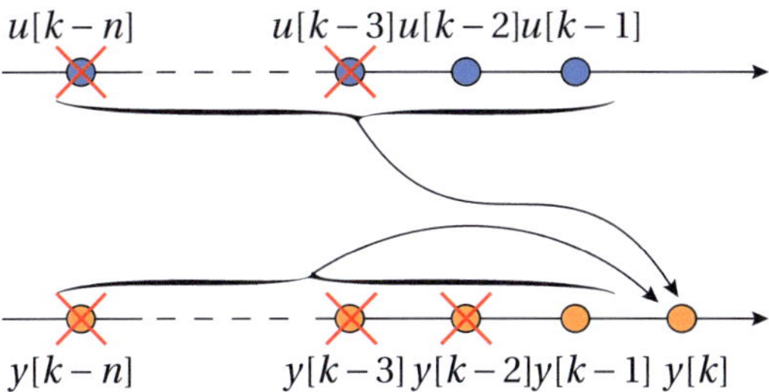

Figure (3.18) Illustration of the basic concept when using SVMs with variable elimination for the detection of causal dependencies. The SVM, containing as input variables past values of $u[k]$ and $y[k]$, eliminates iteratively irrelevant variables based on the prediction accuracy of $y[k]$. This increases the performance of the SVM and deals as test of significance if u causes y. If the found variable subset still contains past values of $u[k]$ it is said that u causes y.

Figure 3.18 sketches the explained basic concept when using SVMs for the detection of causal dependencies. In the following the theoretical foundations of SVMs for regression tasks are reviewed.

Support vector regression　Performing a linear regression, given a training set with $\{x_i, z_i\}_{i=1}^{K}$ with $x_i \in \mathbb{R}^n$ and $z_i \in \mathbb{R}$, means to estimate a function f that fits the training data by minimizing the mean squared error, containing as parameters the normal vector w and the bias b. The regression function is then given by

$$f(x) = \langle w, x \rangle + b, \tag{3.34}$$

while in that case $\langle \cdot, \cdot \rangle$ denotes the scalar product in $\mathbb{R}^n$. For the calculation of $f(x)$, support vectors machines for regression have the objective to find the flattest function of $f(x)$ that has at most a deviation ε (also known as insensitivity zone) from the values z_i for the whole training data set. Flatness in this case means to find a normal vector w which is as small as possible. Taking into account ε, the objective function is written as

$$\min_{w} \frac{1}{2}\|w\|^2 \tag{3.35}$$
$$\text{with } f(x_i) - z_i \leq \varepsilon$$
$$\text{and } z_i - f(x_i) \leq \varepsilon.$$

As the selection of a too small insensitivity zone ε would lead to equations with infeasible constraints, equation 3.35 is extended by so-called slack variables $\xi, \hat{\xi} \in \mathbb{R}_{\geq 0}$ and an additional weighting parameter $C \in \mathbb{R}_{>0}$ [Vap95] defined as

$$\min_{w,\xi,\hat{\xi}} \frac{1}{2}\|w\|^2 + C \sum_{i=1}^{K} (\xi_i + \hat{\xi}_i) \tag{3.36}$$
$$\text{with } f(x_i) - z_i \leq \varepsilon + \xi_i$$
$$\text{and } z_i - f(x_i) \leq \varepsilon + \hat{\xi}_i.$$

For the slack variables $\xi, \hat{\xi}$ commonly a quadratic or a linear loss function is selected, while throughout this thesis the linear loss function will be used. Further details on the selection of loss functions for support vectors machines and their characteristics is given e. g. in [CST00] or [Vog08].

To solve the optimization problem a transformation into its dual form is performed and a Lagrange function is constructed. For further explanations on how to construct this function it is referred to [Van97] and [CST00]. The final result is given as

$$L(w,b,\xi,\hat{\xi},\alpha,\hat{\alpha},\beta,\hat{\beta}) = \frac{1}{2}\|w\|^2 + C \sum_{i=1}^{K} (\xi_i + \hat{\xi}_i) - \sum_{i=1}^{N} (\beta_i \xi_i + \hat{\beta}_i \hat{\xi}_i)$$
$$- \sum_{i=1}^{N} \alpha_i (\varepsilon + \xi_i - z_i + \langle w, x_i \rangle + b) \tag{3.37}$$
$$- \sum_{i=1}^{K} \hat{\alpha}_i (\varepsilon + \hat{x}_i + z_i - \langle w, x_i \rangle - b),$$

while L is the Lagrangian function with $\beta_i, \hat{\beta}_i, \alpha_i, \hat{\alpha}_i \in \mathbb{R}_{\geq 0}^{n}$ being the Lagrange multipliers. In [Van97, CST00] it is stated, that L has a saddle point at the solution, meaning that the partial derivatives with respect to the primal parameters $(w,b,\xi_i,\hat{\xi}_i)$ have to be zero. Therefore, the following four conditions can be calculated as

$$\frac{\partial L}{\partial b} = \sum_{i=1}^{K} (\hat{\alpha}_i - \alpha_i) = 0, \tag{3.38}$$

$$\frac{\partial L}{\partial w} = w - \sum_{i=1}^{K} (\alpha_i - \hat{\alpha}_i) x_i = 0, \tag{3.39}$$

$$\frac{\partial L}{\partial \xi} = C - \alpha_i - \beta_i = 0, \tag{3.40}$$

$$\frac{\partial L}{\partial \hat{\xi}} = C - \hat{\alpha}_i - \hat{\beta}_i = 0. \tag{3.41}$$

Finally, equation 3.39 can be used to calculate the vector w. Setting the result into equation 3.34 leads to the regression function including the Lagrange multipliers $\alpha, \hat{\alpha}$ which are given as

$$f(x) = \sum_{i=1}^{K} (\alpha_i - \hat{\alpha}_i) \langle x_i, x \rangle + b. \tag{3.42}$$

Discussion of $\alpha, \hat{\alpha}$ The prefactors of the support vectors have some important properties depending on the pair (x_i, z_i) of the training data set and the selected parameters ε and C. The derivation of these properties is explained in further detail in [SS04] and [Vog08] and is summarized below.

- $|z_i - f(x_i)| < \varepsilon$: Sample (x_i, z_i) lies inside the insensitivity zone.
  Lagrange multipliers: $\alpha_i = 0, \hat{\alpha}_i = 0$
  Slack variables: $\xi = 0, \hat{\xi} = 0$

- $z_i - f(x_i) - \varepsilon = 0$: Sample (x_i, z_i) lies on the upper edge of the insensitivity zone.
  Lagrange multipliers: $0 < \alpha_i < C, \hat{\alpha}_i = 0$
  Slack variables: $\xi_i = 0, \hat{\xi}_i = 0$

- $f(x_i) - z_i - \varepsilon = 0$: Sample (x_i, z_i) lies on the lower edge of the insensitivity zone.
  Lagrange multipliers: $\alpha_i = 0, 0 < \hat{\alpha}_i < C$
  Slack variables: $\xi_i = 0, \hat{\xi}_i = 0$

- $z_i - f(x_i) - \varepsilon > 0$: Sample (x_i, z_i) lies above the insensitivity zone.
  Lagrange multipliers: $\alpha_i = C, \hat{\alpha}_i = 0$
  Slack variables: $\xi_i = z_i - f(x_i) - \varepsilon, \hat{\xi}_i = 0$

- $f(x_i) - z_i - \varepsilon > 0$: Sample (x_i, z_i) lies below the insensitivity zone.
  Lagrange multipliers: $\alpha_i = 0, \hat{\alpha}_i = C$
  Slack variables: $\xi_i = 0, \hat{\xi}_i = f(x_i) - z_i - \varepsilon$

The list shows some interesting characteristics. As the Lagrange multipliers $\alpha, \hat{\alpha}$ which correspond to the data points lying inside the insensitivity zone all have a value of zero, those data points are not needed for the calculation of the regression function.

Kernels　　One of the main reasons why SVMs are employed is that they have the ability to detect nonlinear dependencies in data. To achieve this, so-called kernel functions are used which are explained in detail e. g. in [CST00, SS01, STC04, Vog08]. The main idea is to map the space of input data $\mathbb{R}^n$ into some feature space $\mathcal{F}$ with a possibly higher dimension by using a nonlinear transformation function Φ. Formally this can be written as

$$\Phi : \mathbb{R}^n \mapsto \mathcal{F} \quad \text{with} \quad x \mapsto \Phi(x). \tag{3.43}$$

The major problem when mapping data into the feature space is, that this approach can become computationally infeasible [SS04]. In that case, the advantage of SVMs is that they solely depend on the calculation of dot products which can be used to reduce the computational complexity significantly. This means that a kernel function, defined as $k(x, x') := \langle \Phi(x), \Phi(x') \rangle$ can be set into equation 3.42 resulting in the final regression function

$$f(x) = \sum_{i=1}^{N} (\alpha_i - \hat{\alpha}_i) k(x_i, x) + b. \tag{3.44}$$

Compared to the case without kernel function this means that the SVM now searches for the flattest function in feature space and no longer in input space [Vog08].

Kernel examples Only functions can be used as kernels that correspond to a dot product in a feature space [SS01]. In the following, some examples of kernel functions are given, while in this thesis solely the Gaussian kernel is used for the detection of causal dependencies. This decision is based on an investigative study made by Rakotomamonjy [Rak07] on using SVMs for variable ranking. He showed that Gaussian kernel work best on most of his proposed benchmark data. Further details about kernels can be found e. g. in [SS01, Vog08, SS04].

- Inhomogeneous polynominal kernel: $k(x,x') = (\langle x, x' \rangle + c)^p$ with $p \in \mathbb{N}, c \geq 0$
  In [VC00, BGV92] it is shown that the mapping into a polynominal feature space can be performed using this kernel. When selecting $p = 1$ and $c = 0$ this corresponds to a linear kernel.

- Sigmoid kernel: $k(x,x') = \tanh\left(\frac{\langle x,x' \rangle + c}{p}\right)$ with $a,r \in \mathbb{R}$
  This kernel corresponds to the activation function of neural networks [DHS01b].

- Gaussian kernel: $k(x,x') = e^{-\frac{||x-x'||}{2\sigma^2}}$ with $\sigma \in \mathbb{R}_{>0}$
  This kernel was first introduced in [ABR64] and is the most widely used kernel for SVMs. It is used throughout this thesis for the reconstruction of the disturbance propagation path and the localization of the root cause of a fault.

3.7.1 Optimizing SVM Parameters

The general behavior of the SVM is strongly dependent on the selected values of the two parameters C and ε for the SVM and σ for the Gaussian kernel. For fitting the parameters to the data set an optimizing technique as well as a loss function need to be defined. In that case, a greedy search algorithm, namely the downhill simplex algorithm, originally developed by Nelder & Mead

[NM65] and as a loss function the squared sum of residuals in combination with the 0.632 bootstrap algorithm [WF05, Efr79] are used. The methods are summarized in the following.

0.632 bootstrap   To evaluate the performance based on the selected SVM parameters, the 0.632 bootstrap algorithm is used which generates a training and a test data set by performing sampling with replacement.

Having a data set with a sample size K, exactly K times instances are taken for training the SVM. As some instances are taken more than one time, other instances are not used for the training set. These samples are later utilized for testing the trained SVM. Having a sufficiently large data set, this means that the possibility for a particular instance of not being picked for training is calculated as $\lim_{K \to \infty} (1 - \frac{1}{K})^K = e^{-1} \approx 0.368$ meaning that 63.2% of the original data set are used for training in which some of the instances occur double and 36.8% are used for testing. By using the residual sum of squares as loss function, the resulting error for the training set E_{training} and for the test set E_{test} are combined into one loss function $V_{0.632}$ given as

$$V_{0.632} = 0.632 \cdot E_{\text{test}} + 0.368 \cdot E_{\text{training}} \,. \tag{3.45}$$

Downhill simplex   Fitting the tuning parameters of the SVM is done by minimizing $V_{0.632}$. The downhill simplex search algorithm used in the present work is one of the most widely used algorithms for optimization. Good descriptions of the algorithm can be found e. g. in [NM65, JRWW98] or [McK98] and are summarized here to give the main points.

The concept is to initialize a simplex in the search space, while a simplex for A parameters is defined as having $A + 1$ elements. For initialization of the SVM the parameters ε, C and σ are chosen. In the next step the selected parameters are used for one element of the simplex and the other A elements are calculated through offsetting always one of the parameters. For each parameter set, the loss function $V_{0.632}$ is calculated and the element with the worst performance is selected. This element is removed and replaced through its reflection on the centroid of the remaining elements. In this way the simplex moves through the search space towards a parameter set representing the minimum of $V_{0.632}$.

Additionally the simplex has the possibility to perform as move reduction, contraction or expansion to find the minimum.

For optimization there are also other methods for support vector machines available, which are e. g. grid-search [HCL03] or gradient-based methods [Gla08]. All of these methods have their assets and drawbacks. Since the suggested method for the detection of causal dependencies only needs to fit the parameters once, the used optimization algorithm is not of crucial importance. Furthermore, as only one SVM is used, even badly fitted parameters, e. g. if the simplex algorithm reaches a local minimum, can be tolerated. This would not be possible when comparing several SVMs. The optimized parameters are denoted as $\varepsilon_{\mathrm{opt}}$, C_{opt} and σ_{opt}.

3.7.2 Variable Selection Methods

Variable selection means to find a subset of relevant variables usually with the aim to have a less complex and more robust model to predict a target variable. Useless noise variables are eliminated and the risk of generating an overfitted model is reduced. Furthermore, lower computational power is needed when the model learning is performed on the reduced variable set. There exists a wide variety of variable selection algorithms, see e. g. [LM98, GE03, KG97], and they can mainly be divided into filter and wrapper methods.

Filter methods are usually computationally cheap as these methods do not necessarily need a learning machine. In that area popular metrics are correlation [DP05] and entropic measures [HFC05] to perform variable selection.

When performing variable selection using a wrapper approach, which is the case when using the SVM, a learning machine has to be trained. Wrappers usually outperform filter methods compared to the prediction error, but are computationally more intensive [ZYSM07]. In that case, the learning machine itself describes a black-box where the prediction performance is used to select the variables. According to Guyon [GE03], when using a wrapper approach besides the learning machine three things need to be defined. They are the strategy on how to search the space of all possible variable subsets, the objective function for the evaluation of the prediction performance and finally which stopping criterion should be used. All three objectives will be explained in the following section.

Detection of causal dependencies   The first step of the proposed algorithm when using variable selection for the detection of cause-effect relationships is to generate a data set containing past sequences of the possible input series $u[k]$ and output series $y[k]$. For both signals the last n values are used to generate an input data set $\Phi_{uy} = \{u[k-1],...,u[k-n],y[k-1],...,y[k-n]\}$ for the SVM and setting $y[k]$ as output value. The SVM is once trained and optimized regarding ε, C and σ on the complete data set. In the next step the variables in Φ_{uy} are ranked in terms of their prediction performance for $y[k]$ using a backward variable elimination method which is explained in the section below. Finally a reduced SVM, containing only a relevant subset variables, is calculated. If the resulting subset contains one or several past values of $u[k]$, a significant causal dependency is found, meaning that u causes y. The time horizon n is estimated as described in section 3.5.2 for the method based on transfer entropy by using a VAR model and the AIC criterion on the output series $y[k]$. In the following, the resulting model order for the SVM is named n_{SVM} and the maximum model order to be estimated is set to $n_{\mathrm{max}} = 10$.

3.7.3 Using Support Vectors for Variable Selection

There has already been several works on how to use SVMs for variable selection. Vapnik [VC00] and Joachims [Joa00] showed that SVMs can be used for variable selection in classification tasks and Chang [CL05] extended this work to use variable selection as well for regression tasks. In both approaches the idea for feature selection is to minimize the Leave-One-Out (LOO) error of the SVM. On a data set with the sample size K the LOO is calculated by using always $K-1$ samples for training the SVM and the remaining data point as a test sample. This all is repeated exactly K times meaning that each sample is used exactly one time for testing. Therefore, the Leave-One-Out error is given as

$$\mathrm{LOO} = \frac{1}{K}\sum_{i=1}^{K} |f(x_i|\{x_1,...,x_k\}\backslash x_i) - z_i|. \tag{3.46}$$

Calculating the LOO has two advantages. First it can be used on small data sets, as always the maximal possible number of training data is used; secondly it is a deterministic method which gives reproducible outcomes.

A disadvantage when calculating the LOO on a large data set, is that it can become computationally expensive.

To overcome this problem Chang [CL05] suggests two methods for the estimation of LOO error bounds for the SVM. One method still contains a quadratic programming problem and the other one contains the inversion of a matrix, which again makes the calculation computationally expensive. As both methods contain the sum of the Lagrange multipliers $\alpha_i, \hat{\alpha}_i$, Rakotomamonjy [Rak07] proposes to use the multipliers directly for variable selection and defines its ranking criterion as

$$G_\alpha(\alpha, \hat{\alpha}) := \sum_{i=1}^{K} (\alpha_i + \hat{\alpha}_i) \propto \text{LOO}. \tag{3.47}$$

In [Rak07] it is demonstrated that for the variable selection this criterion gives similar results as the computationally more expensive methods suggested by Chang. Hence, the approach suggested by Rakotomamonjy will be used for the estimation of the LOO error and consequently for ranking the selected input variables.

Recursive variable elimination   A heuristic algorithm for variable selection is needed as testing all possible variable sets would be computationally infeasible. Therefore, an algorithm originally proposed by Couvreur [CB00] and investigated in further detail by Rakotomamonjy is used. In this approach at first all variables are considered and recursively eliminated by removing the variable which increases the LOO error the least. The concept is to remove temporarily one variable at a time and to calculate $G_\alpha(\alpha, \hat{\alpha})$ using the remaining variables. By comparing the different results, the variable which minimizes $G_\alpha(\alpha, \hat{\alpha})$ is removed. This is done until all variables are ranked. The first removed variable is ranked last and the last removed variable is ranked first. This algorithm is summarized as algorithm 4, when using as input the data set Φ_{uy}.

As result the ranked set of variables stored in $\Phi_{uy}^{\text{ranked}}$ is obtained.

Algorithm 4: Recursive variable ranking, originally proposed by [CB00]

Initialize: SELECTED = $\emptyset$ and VAR = Φ_{uy} ;

while *VAR is not empty* **do**

    for *all variables in VAR* **do**

        Remove temporarily variable i in VAR ;

        Calculate $G_\alpha(\alpha, \hat{\alpha})$;

    end

    SELECTEDVAR = $\arg\min_i G_\alpha(\alpha, \hat{\alpha})$;

    Rank variable: SELECTED = [SELECTEDVAR SELECTED] ;

    Remove variable SELECTEDVAR from VAR ;

end

3.7.4 Significant Subset and Causal Strength

The conducted variable ranking already gives hints for causal dependencies. If past values of $u[k]$ are ranked in high positions this indicates a causal dependency pointing from $u \rightarrow y$. This needs to be covered into a statistical test. Therefore, a relevant subset of the ranked variables $\Phi_{uy}^{\text{ranked}}$ is calculated by means of the prediction performance of $y[k]$. This is done by comparing two SVMs, each containing a different set of variables. The first SVM uses the first ϕ variables in $\Phi_{uy}^{\text{ranked}}$ and the second SVM the first $\phi + 1$ variables for predicting $y[k]$ where $\phi \in \{1,\dots,2n-1\}$. In both cases the parameters are kept to $\varepsilon_{\text{opt}}, C_{\text{opt}}$ and σ_{opt}, which means that SVMs with identical parameters are used.

For the selection of the subset size an F-test is performed on the resulting residual sum of squares of the two SVMs. According to [Joh93] the F-test is in that case calculated as

$$F(K-1, K-1) = (E_1 - E_2) \cdot \frac{K - \phi + 1}{E_2}, \tag{3.48}$$

where E_1 is the residual sum of squares of the first SVM with ϕ variables and E_2 is the residual sum of squares of the second SVM with $\phi + 1$ variables.

The null hypothesis is defined as $H_0 : E_1 = E_2$, meaning that if the null hypothesis cannot be rejected the residual sum of squares do not change significantly and the found subset of variables is set to size ϕ. In other words, the subset of variables, starting with a subset which contains only one variable, is increased in each step by one, until the first time the null hypothesis cannot be rejected. If the found variable subset includes at least one sequence of past values of $u[k]$ it is assumed that u causes y. When performing the F-test, the significance level is set to $\alpha = 0.05$.

Algorithm 5: Summary of the proposed algorithm based on support vector machines

---

1. Estimate the time horizon n_{SVM} using a VAR model and AIC to generate Φ_{uy};

2. Train SVM and fit user selected parameters ε, C, σ using downhill simplex algorithm and check consistency of the SVM using Durbin-Watson statistic;

3. Perform variable selection with $G_\alpha(\alpha, \hat{\alpha})$ and calculate subset;

4. If u is in the subset, set Q^{SVM} as the resulting value of the causal strength $u \to y$;

Causal strength Similar to Granger causality the resulting causal strength is calculated based on the comparison of the squared sum of residuals. The causal strength Q^{SVM} is calculated through a comparison of the two different squared sums of residuals named E_{uy} and E_y. In detail, E_{uy} is calculated using the above explained SVM with the subset of input variables resulting from the initial set Φ_{uy}. For prediction of $y[k]$, the residual sum of squares E_y is calculated by performing the same algorithm only starting with the reduced set $\Phi_y = \{y[k-1],...,y[k-n]\}$ which does not contain the time series $u[k]$ and by using the same parameters $\varepsilon_{\text{opt}}, C_{\text{opt}}$ and σ_{opt}.

Again a tuning parameter β_{SVM} is defined and set to one for the benchmark data. Tuning the parameter is postponed to section 4.2. The resulting value

Q^{SVM} is therefore defined as

$$Q^{\text{SVM}} := \left(1 - \frac{E_{uy}}{E_y}\right)^{\beta_{\text{SVM}}}, \qquad (3.49)$$

with $0 \le Q^{\text{SVM}} \le 1$, where 0 equals no causal dependency and 1 means maximum causal strength. The complete algorithm using support vector machines for the detection of causal dependencies is summarized as algorithm 5, while for the generation of the causal matrix each variable is tested against each other.

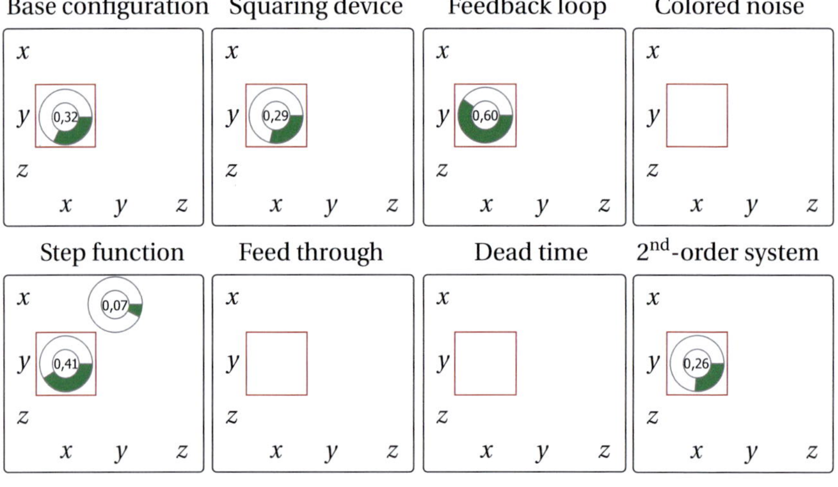

Figure (3.19) Causal matrices represented as doughnut charts when using support vector machines for the detection of the causal direction of the tested benchmark data sets. The red squares represent the a priori known causal dependencies.

3.7.5 Tests with the Benchmarks

In this section the proposed algorithm is tested on the benchmark data sets presented in chapter 3.2. The resulting causal dependencies are outlined in figure 3.19 for each benchmark in terms of a doughnut chart. The initial parameters for the SVM for all benchmarks were set to $C = 100, \varepsilon = 1$ and for the Gaussian kernel $\sigma = 1$. The order of the maximum of the model order is set to $n_{\max} = 10$. Reviewing the results, starting with the **base configuration**, the causal dependency $u \to y$ is correctly found for this benchmark. In that case, the dependency yields a causal strength of $Q^{\text{SVM}} = 0.32$.

The causal dependency $u \to y$ in the nonlinear benchmark data set consisting of the **squaring device** is correctly detected as well. This reveals the difference of the SVM compared to Granger causality, as due to the fact that a Gaussian kernel is used, even nonlinear causal dependencies in the data can be found.

The **feedback loop** has no negative impact on the SVM and the causal dependency in the benchmark is correctly detected.

Having the benchmark with **colored noise**, no causal dependencies are found in the data set. The reason is that the output series $y[k]$ contains already too much information about itself. As a consequence the feature selection algorithm removes the possible input variable u from the subset.

The benchmark consisting of white noise with a superposed **step function** as input implies a significant cause-effect relationship $u \to y$ and $y \to u$ while $u \to y$ results in a larger causal strength than $u \to y$. Like for GC, this shows again the advantage when using the causal strength as a quantitave measure. Detecting causal dependencies depending only on significant tests would result in this case in an equal intercausal relationship, meaning that the two variables would have the same causal influence onto each other.

No causal dependency could be detected regarding the benchmarks containing the **feed through** and the **dead time**. Like the Granger causality the SVM can only cover dead times through its model order n. Since for the dead time a model order of $n = 21$ ($T_{\text{d}} = 2\,\text{s}$, $T_{\text{s}} = 0.1\,\text{s}$) is needed, but as shown in figure 3.20 the model order of the VAR model is estimated to 4, no cause-effect relationship is found. The reason is that the model order is entirely estimated through the output signal $y(t)$ which does not contain any information about the dead time of the system.

In terms of the feed through too little causal information is contained in u, so that the delayed vectors of u are not in the estimated subset.

For the last benchmark data, which contains the **2nd-order system**, the SVM finds the correct causal dependency resulting in a slightly lower causal strength compared to the base configuration.

Furthermore, the causal matrices for all benchmarks show that the added time series $z[k]$ from another process is always correctly detected as noise variable. This means that in all cases the variable selection algorithm correctly sorted out this variable from the subset.

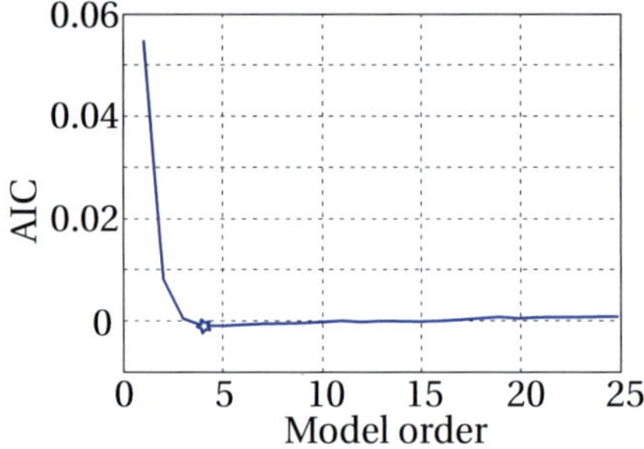

Figure (3.20) Resulting values of the AIC for the SVM when having a large dead time in the system. The star marks the estimated model order of the SVM.

4

Method Selection

**The last chapter has illustrated that each method has its typical character-
istics when calculating causal dependencies. Hence, this chapter starts by
summarizing the outcome from the benchmark data and by giving advice
which methods to use depending on the system dynamics and the expected
disturbances. Thereafter, it is explained how the methods can be combined
to one resulting causal matrix and how a priori knowledge in terms of known
cause-effect dependencies can be used. The chapter finishes by testing the
methods on generated data from a simulated continuous stirred tank reactor.**

4.1 Selection depending on Process Characteristics

In chapter 3 several methods have been proposed and tested on eight different benchmark data sets. The results illustrated that each method has its
advantages and drawbacks depending on the type of the disturbance and the
process characteristics. This means that if there is a priori information about
the underlying system available, certain methods can be selected or eliminated
beforehand.

Table 4.1 summarizes the results from the benchmark data sets. The checkmarks describe if it is possible for a method to find the causal dependency in

the benchmark. Checkmarks being set in brackets indicate that the method needs to be applied with attention, as causal dependencies were not doubtlessly detected.

	CCF	TE	GC	SVM
Base configuration	√	√	√	√
Squaring device	–	√	–	√
Feedback loop	√	√	√	√
Colored noise	–	–	√	–
Step function	–	√	(√)	(√)
Feed through	–	–	(√)	–
Dead time	√	√	–	–
2nd-order system	√	√	√	√

Table (4.1) Summary of the results from the investigated benchmark data sets. A checkmark describes that the proposed method is able to detect the causal dependency correctly. Checkmarks being set in brackets indicate that the method needs to be applied with attention as causal dependencies were not doubtlessly detected.

Under the assumption that the results of the benchmarks can be generalized it is possible to use this table to generate a sunburst graph for the selection of the best method(s) corresponding to an underlying system. The graph is illustrated in figure 4.1.

In the case of having a linear system or a system structure that can be linearized adequately while having no or only short dead times, all methods can be used for the detection of causal dependencies. As Granger causality and support vector machines can only cover delay times up to the selected internal model order n, for large dead times, the cross-correlation function or transfer entropy should be selected as they cover dead times through a time-shifting parameter λ. If the process is nonlinear, the support vector machine or transfer entropy need to be used as cross-correlation and Granger causality need linear similarities in the process signals. When dealing with large dead times and having additionally a nonlinear behavior only the transfer entropy can be applied for the detection of causal dependencies.

When selecting methods based on the sunburst graph it has to be taken into account that this selection approach does not cover the disturbance characteristics. In terms of the benchmark data, narrow-banded disturbances and the impact of a set-point change have been investigated. From table 4.1 it can be concluded not to use the CCF if it is expected that the disturbance implies a change of the stationary value. Having a narrow-banded disturbance, Granger causality should be selected as it is the only method that could detect this cause-effect relationship in the benchmark data set. Still, all methods are negatively affected in their performance by narrow-banded signals. This is illustrated at the end of the chapter using data from the continuous stirred tank reactor. In addition, it is always possible to combine several methods, e. g. using the cross-correlation for the estimation of the dead time and afterwards the SVM for the detection of the essential cause-effect relationships. This is not represented in the graph but becomes obvious from the results presented in table 4.1.

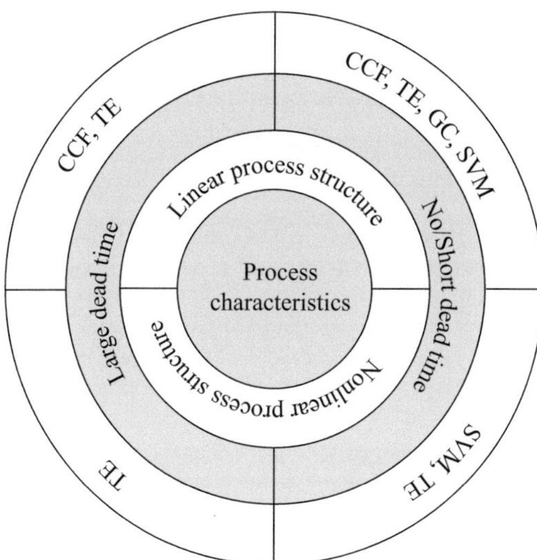

Figure (4.1) Sunburst graph summarizing which method should be selected for fault localization depending on the characteristics of the available process structure.

4.2 Combining the Methods

As mentioned, the results of the benchmark data sets have shown that each method has its advantages and disadvantages. If the process characteristics or the expected type of disturbance are well known the appropriate method(s) can be select. When there is no or only little knowledge of the system at hand, the approach proposed in the following is to combine the methods by merging them into one resulting causal matrix. Possible drawbacks of one method can then be covered through the advantages from other methods and more robust results regarding the found disturbance propagation path can be obtained. Nevertheless, the outcome of the benchmark data indicates that the mean value of the causal strengths for the significant causal dependencies for each method differs. Hence, in section 4.2.1 a design concept is proposed which shows how the methods can be combined. This is necessary to avoid that one method dominates the resulting combined causal matrix. If some cause-effect relationships are already known beforehand an approach for weighting the methods can be used. This is explained in section 4.2.2.

	CCF $\beta_{CCF} = 1$		TE $\beta_{TE} = 1$		GC $\beta_{GC} = 1$		SVM $\beta_{SVM} = 1$	
	$u \to y$	$y \to u$	$u \to y$	$y \to u$	$u \to y$	$y \to u$	$u \to y$	$y \to u$
1st-order system	0.86	–	0.04	–	0.49	–	0.32	–
Feedback loop	0.91	–	0.11	–	0.56	–	0.60	–
2nd-order system	0.85	–	0.04	–	0.50	–	0.26	–
$\overline{Q}$	0.87		0.06		0.52		0.34	

Table (4.2) Resulting causal strengths for the three benchmarks that were detected correctly by all methods while keeping β for all methods at the value of one. Q^{CCF} leads to the highest and Q^{TE} to the lowest mean value $\overline{Q}$ for the causal strength.

4.2.1 Balancing Causal Matrices

Each method leads to a causal matrix $Q \in \mathbb{R}^{r \times r}$ with r process variables and the causal strength $q_{X_i \to X_j} \in [0,1]$. For all methods, $q_{X_i \to X_j}$ is defined as a contin-uous heuristic measure which describes the causal impact one process variable has onto another and increases monotholically with an increasing causal strength. In the following Q denotes the short-term for any of the resulting matrices $Q^{\text{CCF}}, Q^{\text{TE}}, Q^{\text{GC}}$ or Q^{SVM}. As one of the main results from chapter 3 it is stated that each method yields a different mean of the causal strengths for the benchmark data sets. This becomes obvious when comparing the three benchmarks which could be found by using any of the proposed methods, namely the 1st-order system, the feedback loop and the 2nd-order system. The resulting causal strengths are summarized in table 4.2. In that case the CCF implies the highest and the TE the lowest values.

To balance the methods to each other, the introduced exponential fitting parameters $\beta^{\text{CCF}}, \beta^{\text{TE}}, \beta^{\text{GC}}, \beta^{\text{SVM}} \in [0, \infty)$ are used. In the following, as a short-term β is applied to describe the set of fitting parameters. As mentioned, the absolute values of the causal matrices of the four methods are not comparable directly, since each method uses a different mathematical approach to calculate the cause-effect dependencies. Still, each found causal relationship for itself is relevant as all methods work with statistical significance tests. Hence, by selecting appropriate values for the different fitting parameters β, this means that also the resulting causal strengths from the different methods can be compared up to a certain point.

As shown in chapter 3, for balancing the methods an exponential fitting is selected. This assures that all causal strengths of the balanced matrices result in values between zero and one. For the investigation of the methods on the benchmark data, β was set in all cases to one.

Balancing matrices depending on the data set   When balancing the matrices of the different methods to each other, one has to deal with the problem that there exists an almost unlimited variety of disturbances. This means that it is not possible to calculate a fixed β for the methods for all possible disturbance that can possibly occur in a process. In other words, the tuning parameter β needs to be adapted for each method depending on the analyzed data set.

Therefore, if no a priori knowledge is available, the proposed design approach is based on the assumption that on the average all methods will work equally well on the data set. In that case, equally well implies that for the found significant causal dependencies all methods give the same mean value. Hence, the value of the parameter β is fitted for the different methods in a way that the causal matrices Q^{CCF}, Q^{TE}, Q^{GC} and Q^{SVM} give the same mean values defined as $\overline{Q}$ for the detected cause-effect relationships. Throughout this thesis the value for $\overline{Q}$ is set to 0.5.

Under the assumption that all data from the benchmarks is generated from one data set, table 4.3 gives the results of the balanced causal matrices. The resulting β for each method is given in table 4.3 as well. For a better comparison of the methods, figure 4.2 illustrates the results from table 4.3 for the benchmark data in terms of bar charts.

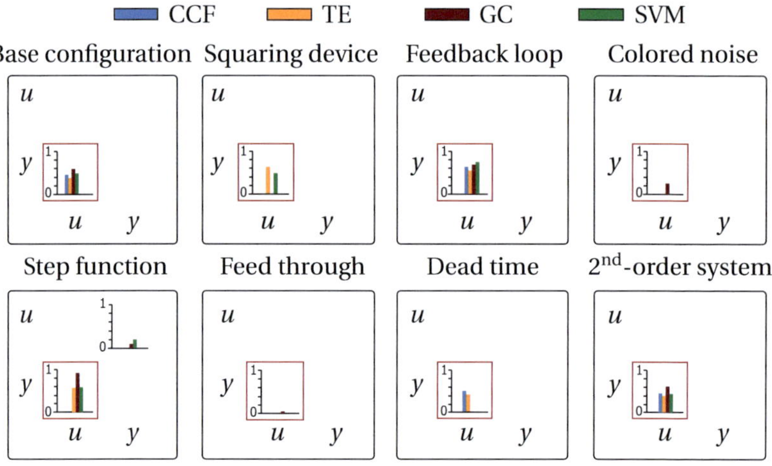

Figure (4.2) Resulting bar charts of the balanced causal matrices for the benchmark data sets. The red squares represent the known causal dependency in the data. The illustration shows that each method has its assets and drawbacks depending on the process characteristics and the type of disturbances.

Calculating combined causal matrix   When having no a priori knowledge in terms of process knowledge at hand it has to be assumed that the causal matrix from each method contains the correct disturbance propagation path. Hence, the calculation of the combined causal matrix Q_{fus} (fus = fused) is performed by taking the mean over all four balanced causal matrices defined as

$$Q_{\text{fus}} := \frac{Q^{\text{CCF}} + Q^{\text{TE}} + Q^{\text{GC}} + Q^{\text{SVM}}}{4}. \tag{4.1}$$

Regarding the laboratory plant presented in chapter 5, for all generated faults, the matrices of the different methods are balanced to each other and equation 4.1 is used to calculate the resulting causal matrix for the root cause priority list.

	CCF $\beta_{\text{CCF}} = 5$		TE $\beta_{\text{TE}} = 0.29$		GC $\beta_{\text{GC}} = 0.7$		SVM $\beta_{\text{SVM}} = 0.62$	
	$u \to y$	$y \to u$	$u \to y$	$y \to u$	$u \to y$	$y \to u$	$u \to y$	$y \to u$
1st-order system	0.47	–	0.39	–	0.61	–	0.50	–
Squaring device	–	–	0.61	–	–	–	0.47	–
Feedback loop	0.62	–	0.53	–	0.67	–	0.73	–
Colored noise	–	–	–	–	0.25	–	–	–
Step function	–	–	0.57	–	0.91	0.09	0.58	0.19
Feed through	–	–	–	–	0.04	–	–	–
Time delay	0.47	–	0.39	–	–	–	–	–
2nd-order system	0.45	–	0.39	–	0.61	–	0.44	–
$\overline{Q}$	0.5		0.5		0.5		0.5	

Table (4.3) Causal strengths for the benchmark data sets after balancing the causal matrices. It is assumed that each method works equally well on the benchmark data sets and $\overline{Q}$ is set to 0.5. The resulting β for each method is given as well.

4.2.2 Including knowledge in terms of known cause-effect dependencies

In section 4.1 it was shown how the different methods can be selected depending on known process and disturbance characteristics. Another way to include knowledge when calculating the combined causal matrix is if some causal dependencies are already known beforehand.

In that case, the main approach is to change equation 4.1 under the assumption that methods giving good results on the known causal dependencies also yield good results on the whole data set. Therefore, the methods working well on these dependencies are weighted stronger in Q_{fus} than methods not detecting the known cause-effect relationships. The approach uses some ideas which are also used for boosting in classification tasks which will not be explained in further detail. For further literature about boosting it is referred to [MR03] or [SF12].

To explain the approach, the proposed methods CCF, TE, GC and SVM, are defined as being a set of four weak classifiers $h_1(x),\ldots,h_4(x)$, which can be combined to result into one strong classifier. The classifiers depend on a given data set $\{x_i,z_i\}_{i=1}^{K}$ with $x \in X$ being some feature space and $z \in \{-1,1\}$ defining two classes. The value of K describes the available sample size. In the case of causal discovery the two classes are used to define if a causal dependency is existent or not. In general, the constructed strong classifier leads to better results than using the different weak classifiers separately on the same data set.

In principle all methods are unsupervised learners, meaning that no training set is available for testing if the methods work well on a data set or not. Still, if some process knowledge in terms known cause-effect relationships is available, the information can be used to generate a supervised training data set to weight the methods. In that case, it is defined that $z_i = 1$ if an existing causal dependency is known a priori and $z_i = -1$ if a priori it is known that no causal dependency exists between the two variables. Therefore, regarding the used methods the resulting causal strengths $q_{X_i \to X_j}$ need to be transformed into binary values meaning that

$$h_n(x_i) = 1 \quad \text{if} \quad q_{X_i \to X_j} \quad \text{is significant,} \tag{4.2}$$

$$h_n(x_i) = -1 \quad \text{if} \quad q_{X_i \to X_j} \quad \text{is not significant,} \tag{4.3}$$

where the index n describes the applied method. In other words, the generated causal matrices are used to indicate if the significance tests have succeeded or failed. For each method the error rate w_n is then defined as

$$w_n := \sum_{i:h_n(x_i) \neq z_i}^{K} \frac{1}{K}, \tag{4.4}$$

with $w_n \in [0,1]$. In a next step, the weight $\alpha_n \in \mathbb{R}$ for each method (or classifier) can be calculated. In the following, the weight is defined as

$$\alpha_n := 0.5 \log \frac{1 - w_n}{w_n}. \tag{4.5}$$

By calculating α_n for each method and including this into equation 4.1, the causal matrix $Q_{\text{fus}}^{\text{pk}}$ (pk = process knowledge) can finally be written as

$$Q_{\text{fus}}^{\text{pk}} := \frac{\alpha_1 Q^{\text{CCF}} + \alpha_2 Q^{\text{TE}} + \alpha_3 Q^{\text{GC}} + \alpha_4 Q^{\text{SVM}}}{\sum_{n=1}^{N} \alpha_n}. \tag{4.6}$$

How the proposed approach can help to increase the probability for the detection of the root cause of a disturbance is demonstrated below by using data from a simulation of the continuous stirred tank reactor.

4.3 Continuous Stirred Tank Reactor

The continuous stirred tank reactor (cstr) was introduced in section 2.4 for the detection of causal structures when using probabilistic measures. In the following, a model of this plant is used for fault localization and the calculation of the disturbance propagation path using the suggested methods with and without a priori knowledge. In total, four different scenarios are investigated while in all scenarios the fluid temperature ϑ_{fl} and the educt concentration c_{in} are both superposed with noise to simulate the disturbance. For the first scenario ϑ_{fl} is superposed with zero-mean white noise having $\mathcal{N}(0,3\,\text{K}^2)$ and c_{in} is superposed with white noise having $\mathcal{N}(0,0.1\,(\text{mol}/\text{L})^2)$. In the other three scenarios colored noise is used to superpose ϑ_{fl} and c_{in}. Therefore, the white noise is filtered beforehand using a 1st-order low-pass filter.

For the second scenario the time constant is set to $T = 1\,\text{s}$, for the third scenario to $T = 5\,\text{s}$ and for the last scenario to $T = 10\,\text{s}$ to generate the disturbance. In terms of the low-pass filter this means, that the higher the time constant is selected, the more narrow-banded is the signal.

Referring to the benchmarks (see e.g. figure 4.2) the results indicate that all methods face problems when trying to detect causal dependencies when having disturbances with a limited bandwidth.

In the following, the impact of disturbances with limited bandwidth are investigated with and without a priori knowledge using receiver operating characteristics.

Compared to the benchmark data another difference is that there are now two causes, namely ϑ_{fl} and c_{in}, having an impact on the resulting measured concentrations c_A, c_B, c_C. For data acquisition, each scenario is simulated with a sample time of $T_s = 0.1\,$s and to avoid numerical issues $K = 10000$ samples are used. An extract of the data set used in the first scenario is illustrated in figure 4.3.

4.3.1 Causal Matrix Using White Noise as Disturbance

The resulting causal matrix for the scenario with superposed white noise is illustrated in figure 4.4 while the red squares indicate the expected causal dependencies. For all methods the mean causal matrix $\overline{Q}$ is set to 0.5, which results for the fitting parameters in $\beta_{CCF} = 0.51$, $\beta_{TE} = 0.11$, $\beta_{GC} = 0.59$ and $\beta_{SVM} = 0.28$. Compared to the values of β for the benchmark data in table 4.3, this gives different estimations. This again illustrates, that the different β need to be estimated each time depending on the underlying data set.

The causal matrix in figure 4.4 indicates that all cause-effect dependencies are correctly detected. Additionally, the SVM detects a false causal dependency pointing from c_C to c_B. The disturbance propagation graph visualizes the resulting causal dependencies. Since the suggested methods do not differentiate between direct and indirect causal dependencies this corresponds to the real causal structure from the differential equations given in equation 2.10 with a wrong weak causal dependency pointing from c_C to c_B. c_{in} and ϑ_{fl} have an impact on all three concentrations c_A, c_B and c_C, the concentration c_A has an impact on c_B and the concentration c_B has an impact on c_C. Transforming Q_{fus} into a root cause priority list, which was introduced in section 3.3, is given in table 4.4.

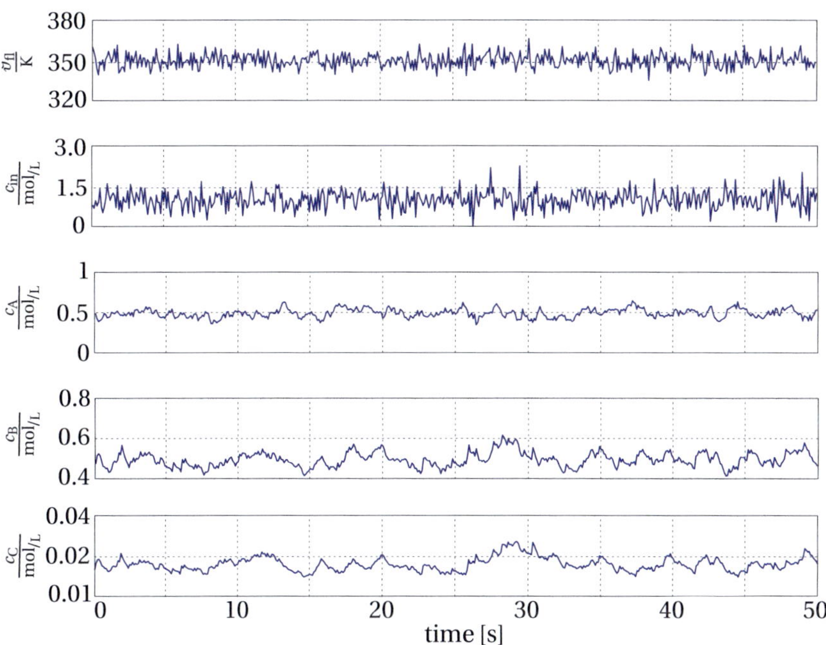

Figure (4.3) Measurement data acquired from a simulation of the continuous stirred tank reactor used for fault localization and the calculation of the disturbance propagation paths. ϑ_{fl} and c_{in} are superposed with noisy data to generate disturbances which propagate through the process.

Rank	Process variable	RC
1	ϑ_{fl}	2.05
2	c_{in}	1.20
3	c_A	0.48
4	c_B	0.20
5	c_C	0.12

Table (4.4) Root cause priority list calculated from the causal matrix in figure 4.4. The two first ranked variables, namely ϑ_{fl} and c_{in} are the root causes of the disturbance. Additionally, position three to five represent the reaction chain c_A, c_B, c_C of the continuous stirred tank reactor.

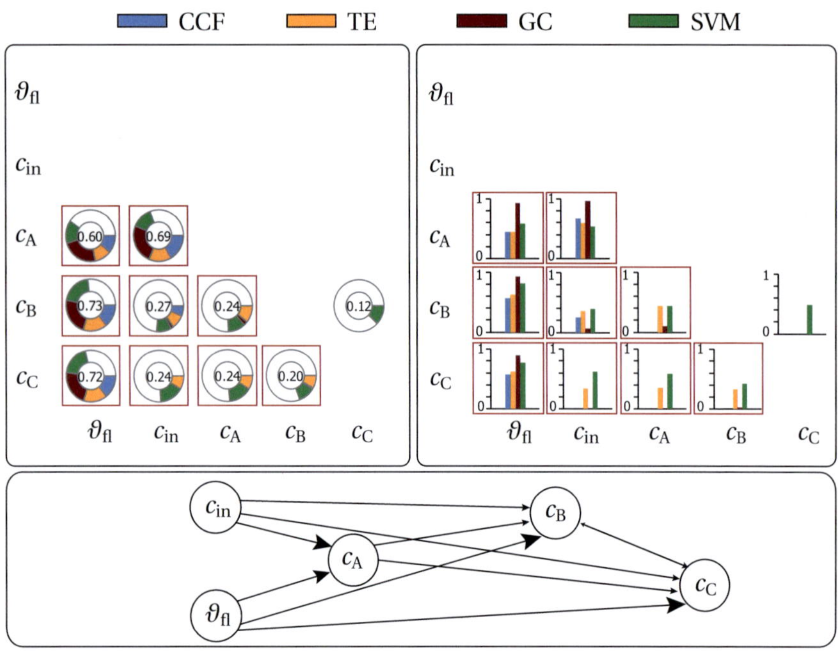

Figure (4.4) Causal matrices when combing all methods while having white noise as disturbance. The left plot shows the combined causal matrix, the right plot illustrates for comparison each method in terms of a bar chart, where the red squares indicate the known causal dependencies. The resulting disturbance propagation paths are visualized in the lower plot.

This shows that the list follows exactly the propagation of the disturbance in the continuous stirred tank reactor. Most importantly ϑ_{fl} has been correctly detected as being the root cause and c_{in} is ranked on position two. Analyzing the bar chart in figure 4.4 illustrates that all methods detect a large causal strength of ϑ_{fl} pointing towards the other process variables. This becomes obvious when taking into account the underlying differential equations of the cstr (see equation 2.10) as the temperature has a direct impact on all three concentrations. Furthermore, the result indicates that the nonlinearity implied in the exponential function has been correctly fitted by GC and CCF as they need linear similarities in the process signals.

Another high causal strength has been found from $c_{in} \rightarrow c_A$. This can also be explained through the differential equations as c_{in} has a direct impact on c_A. The relationship $c_{in} \rightarrow c_B$ is the only indirect causal dependency detected by all methods, $c_A \rightarrow c_B$ and $c_B \rightarrow c_C$ are detected by GC, TE and the SVM. The cause-effect dependency $c_{in} \rightarrow c_C$ and $c_A \rightarrow c_C$ has been found by TE and the SVM. As the disturbance travels through the process this means that it is low-pass filtered. Therefore, the causal relations detected later in the causal chain result in lower causal strengths and significant dependencies are detected with fewer methods.

As mentioned, in the other three investigated scenarios the disturbance is already low-pass filtered beforehand meaning that the disturbance propagation paths will be less obvious. A large time constant, which defines the bandwidth of the disturbance, makes it more difficult for the methods to detect the disturbance propagation paths and the root cause. The results of all four scenarios are represented in terms of receiver operating characteristics (ROC).

4.3.2 ROC for the Different Noise Scenarios

To calculate the probability for the correct detection of the root cause for each of the four noise scenarios the mean is taken over $N = 200$ runs and $K = 10000$ samples are used for simulation. The mean causal matrix $\overline{Q}$ is set to 0.5 for all methods. As the system contains two root causes, two receiver operating characteristics are generated for each scenario. Therefore, the first ROC represents the input variable ϑ_{fl} and the second one the input variable c_{in}. To investigate the advantage when having additional a priori knowledge about the known cause-effect dependencies at hand and by following the approach presented in section 4.2.2 it is assumed that it is known that c_C is the resulting product of the otherwise not known chemical reaction. Regarding the approach when having a priori knowledge in terms of cause-effect dependencies at hand this means that it is known that the two dependencies $c_A \rightarrow c_C$ and $c_B \rightarrow c_C$ exist and that there is no causal dependency pointing from c_C to c_A and from c_C to c_B.

The resulting ROCs for all four scenarios are illustrated in figure 4.5. The receiver operating characteristic without process knowledge is described through Q_{fus} and the one with process knowledge is given through Q_{fus}^{pk}. The scenario with white noise as a disturbance leads to the theoretical optimum for both root causes. The root cause priority list gives in all simulated runs the correctly ranked variables with ϑ_{fl} on the first position and c_{in} being ranked second. Q_{fus} and $Q_{\mathrm{fus}}^{\mathrm{pk}}$ both result in the same ROC and the two functions superpose.

The negative impact of colored noise becomes already visible in the ROC when filtering with a time constant of $T = 1\,\mathrm{s}$. In nearly 80 % of the runs ϑ_{fl} has been set correctly on the first position in the root cause priority list and in 20 % it has been set on position two. When investigating the first two variables in the list, ϑ_{fl} and c_{in} are detected in 100 % of the cases which means that in this scenario in some runs ϑ_{fl} was ranked second and c_{in} first. A priori knowledge about the reaction scheme has in this case almost no impact as the functions resulting from Q_{fus} and $Q_{\mathrm{fus}}^{\mathrm{pk}}$ differ only slightly.

Having colored noise with a time constant $T = 5\,\mathrm{s}$ illustrates the positive effect when including a priori knowledge in terms of known cause-effect dependencies. The function regarding ϑ_{fl} generated from $Q_{\mathrm{fus}}^{\mathrm{pk}}$ results in a higher probability of detecting the root cause when selecting the first ranked variable. Regarding c_{in}, when investigating the second variable, the resulting function from $Q_{\mathrm{fus}}^{\mathrm{pk}}$ is also above Q_{fus}.

The last scenario with a time constant $T = 10\,\mathrm{s}$ illustrates the negative impact of narrow-banded noise and again the positive effect when having a priori knowledge at hand. For ϑ_{fl} the ROC generated from $Q_{\mathrm{fus}}^{\mathrm{pk}}$ is significantly higher than Q_{fus} when selecting the first two ranked variables. When selecting the first three ranked variables ϑ_{fl} is detected in around 80 % of the cases without and in nearly 90 % with process knowledge. The generated receiver operating characteristic for c_{in} from Q_{fus} and $Q_{\mathrm{fus}}^{\mathrm{pk}}$ is in both cases close to a random selection of the variables. In other words, having this narrow-banded noise as input signal c_{in} can no longer be detected as being one of the root causes of the disturbance.

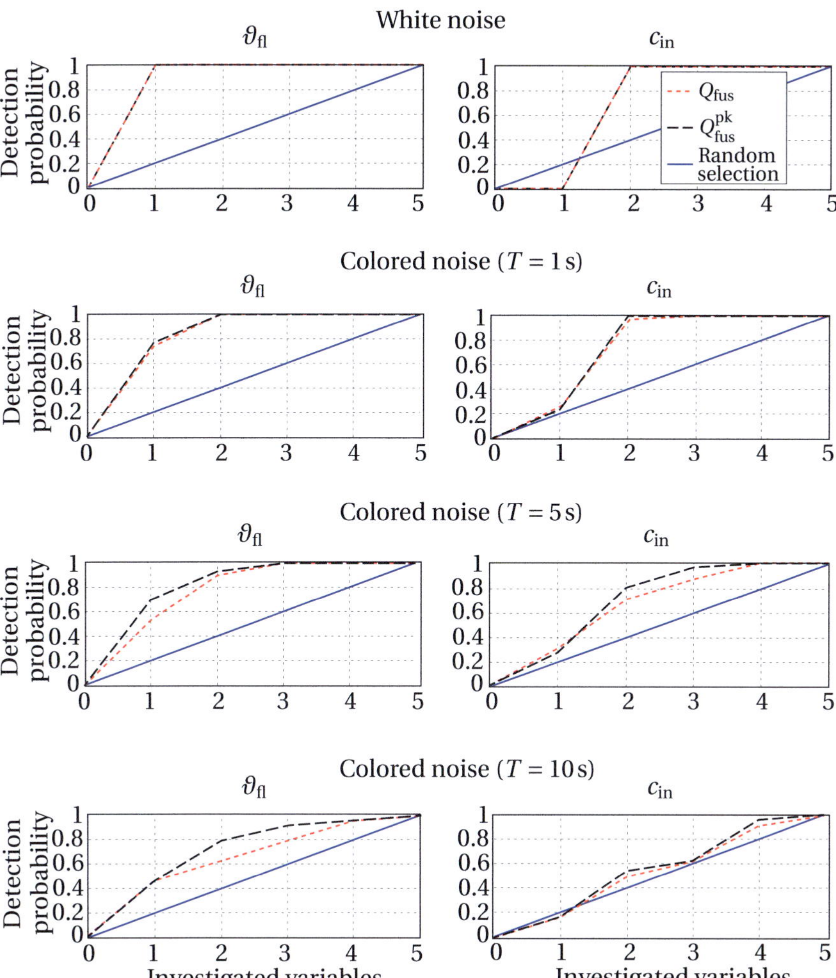

Figure (4.5) Receiver operating characteristics for the continuous stirred tank reactor. The graphics show the impact of the different types of noise with limited bandwidth in the data and when having the advantage of process knowledge compared to an averaging of the methods.

5

Fault propagation on Laboratory Plant

In chapter 4 the methods have been only tested on simulated data. Therefore, this chapter is dedicated to investigate the behavior of the proposed methods on generated faults on a laboratory plant. In the first sections the characteristics of the plant are described and the conducted faults are explained in detail. The other sections are used to investigate the behavior of each method depending on the respective fault and illustrate the advantages when combining the different methods into one resulting causal matrix for the calculation of the disturbance propagation path.

5.1 Setup of the Laboratory Plant

In this chapter the methods developed in chapter 3 are tested for the detection of different disturbance propagation paths using the setup of an experimental laboratory plant. A photo of the plant is given in figure 5.1 and a schematic drawing containing the pertaining process variables is illustrated in figure 5.2. To activate the process, a pump, which is positioned on the lower side of the plant, is set into feed-forward control to transfer water into the ball-shaped upper tank. From the upper tank the water passes several measurement devices before flowing into a lower cylindrical tank. Finally, the water flows from the lower tank back to the pump and closes the water cycle. As process variables between the two tanks the pressure in mbar and the flow in $^{m^3}/_{min}$ are measured. The pressure sensor is positioned directly below the upper tank meaning that it measures the hydrodynamic and hydrostatic pressure of the water in the tank. The flow meter is positioned below the upper tank and behind the pressure sensor. A valve is placed between flow meter and lower tank which can be set into a specific position to control the water flow to the lower tank. The position of the valve is measured in percent while 0 % means that it is closed and 100 % that it is completely open. In addition, in the lower tank the filling level is measured by a level sensor. Like for the valve, the feeding rate of the pump is provided in percent, while 0 % means that the pump is off and 100 % means that it runs in full power. The data is sampled with $T_s = 2\,\mathrm{s}$.

Figure (5.1) Experimental setup for analyzing the disturbance propagation on a laboratory plant. Water is pumped into an upper tank and flows back to a lower tank passing several process devices. Figure 5.2 gives a schematic drawing of the setup.

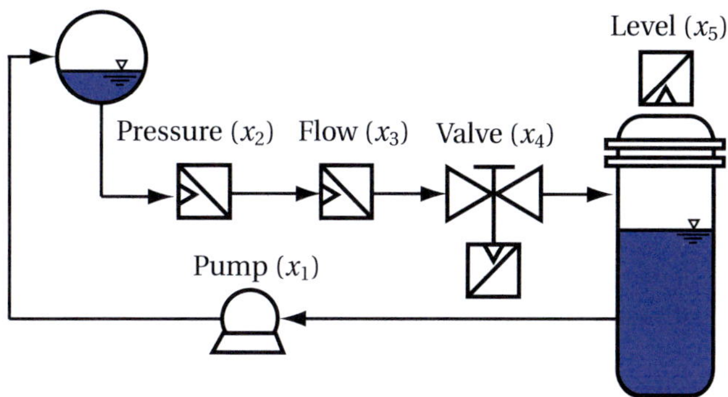

Figure (5.2) Schematic drawing of the laboratory plant with utilized process equipment for analysis. The setup is used to test the suggested methods for the detection of several different root causes applied to the plant.

As stated in chapter 2, cause-effect dependencies in data can only be detected if there is a time delay or a dead time between two process variables. Regarding the laboratory plant, both water tanks represent 1st-order systems (for further description see e. g. [Lun10]) meaning that there is a time delay and that causal dependencies between the pump and the pressure sensor respectively the flow meter can be detected. Furthermore, the tube between the two tanks has a length of approximately one meter meaning that there is a significant dead time between the pump, pressure sensor and flow meter.

To increase the readability for the generated causal matrices a coding is used which follows the water flow in feed-forward control. In detail, the feeding rate of the pump is represented as x_1, the pressure as x_2, the flow as x_3 the valve opening as x_4 and the filling level as x_5. The acquired measurement data when the process is running in normal mode is illustrated in figure 5.3. In normal mode the power of the pump is set to 50 % and the valve is opened by 50 %. The acquired data shows that flow, pressure and level sensors measure only small fluctuations.

5.2 Applied Faults

In total, there are six faults applied to the process to investigate the performance of the proposed methods. The first three experiments are related to real faults on a plant, the fourth and the fifth experiment are more of academic interest. The last experiment covers the trivial case where a fault doesn't lead to a plant-wide disturbance. All experiments have in common that they aim to investigate if it is possible to localize the root cause of the fault correctly and how many samples are needed to give reliable results.

Investigating how many samples are needed is important, as this indicates how much time is needed to detect the fault reliably using the proposed methods. This is represented in terms of receiver operating characteristics (see section 3.3.2) for each experiment, while each function is calculated for different lengths of the acquired measurement data.

Besides finding the root cause of the disturbance, the methods are used to reconstruct the disturbance propagation path. The experiments are described in detail below.

- **Experiment 1: Pump with loose electricity connection** The first experiment addresses the impact of a loose electricity connection in the pump. To generate the fault during data acquisition the pump is randomly switched on and off. When the pump stops, no water flows in the upper tank meaning that the hydraulic pressure is reduced. Switching the pump on means that water is transferred again into the upper tank re-establishing the stationary process behavior. While performing the experiment the valve is kept half-opened.

- **Experiment 2: Valve air pressure leak** The valve in the laboratory plant uses pressurized air for positioning. In this experiment it is assumed that the connection cable from the compressor is not correctly attached to the valve and the valve sometimes closes due to a pressure leak. While the experiment is performed, the pump is set to 50 % of its maximum feeding rate to provide a constant flow from the pump to the upper tank. When the valve closes, the pump still transfers water into the upper tank,

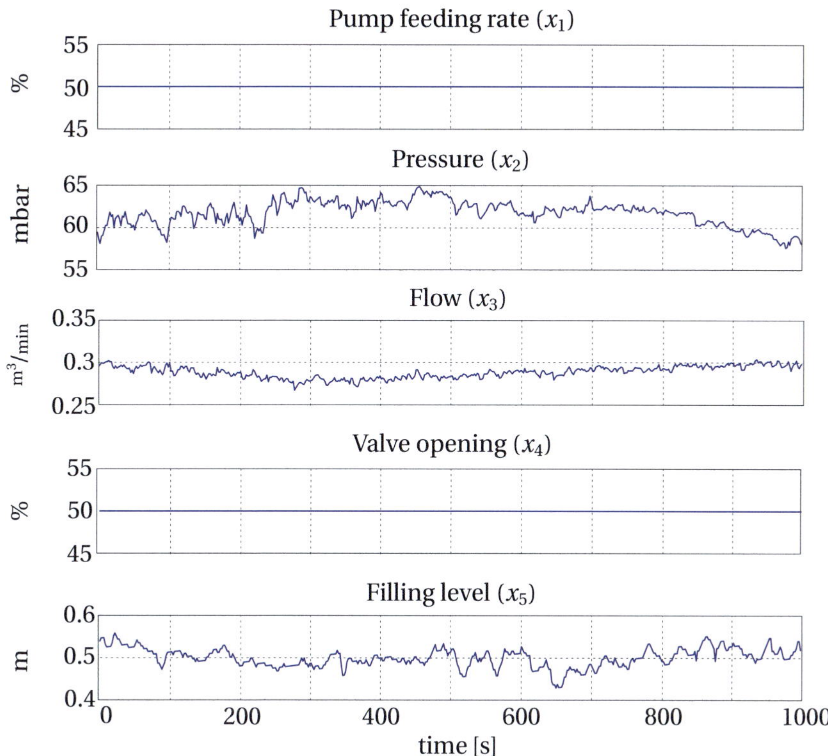

Figure (5.3) Process data from the laboratory plant when running in normal mode without faults. The process variables measure only small fluctuations, pump and valve are set to 50 %.

meaning that the hydraulic pressure increases. At the same time no water flows into the lower tank, meaning that the water level in the lower tank is reduced.

- **Experiment 3: Fault in pump and consecutive fault in the valve** This experiment is a combination of the first two applied faults in the process. In that case the pump and the valve are opened and closed to simulate a fault, while the valve follows the behavior of the pump with a 15 s delay. As there are now two faults acting on the plant, the calculation of the root cause priority list should result in ranking the pump on position 1. The pump should be the root cause, as the valve follows the behavior of the pump. Furthermore, the suggested methods should detect a cause-effect relationship pointing from the pump to the valve.

- **Experiment 4: Oscillating pump with loose electricity connection** In this experiment it is assumed that the flow rate of the pump follows an oscillating waveform as normal behavior. During the data acquisition the pump is stopped twice for a short time and afterwards retakes its oscillating behavior. As the oscillation is cyclic, no cause-effect dependency can be drawn from it. In other words the methods can detect the pump as the root cause only by analyzing the information conducted from the stopping of the pump.

- **Experiment 5: Tube clogging** In this experiment it is assumed that dirt in the tube reduces the water flow from the upper to the lower tank. To simulate this fault, a stop cock, which is positioned between the flow meter and the valve, is partially closed. After approximately six minutes it is reopened. The pressure and the flow meter should detect the disturbance first, as they are located next to it. It is expected that the level meter detects the fault later, as the lower tank acts like a low-pass filter. The valve and pump are both set on 50 % during the experiment.

- **Experiment 6: Broken level sensor** In this experiment it is assumed that the level sensor, positioned on the lower tank, is broken. To generate this fault, the measurement device is moved up and down randomly. Since the sensor is only used for measuring the filling level and is not an acting variable, this fault addresses the trivial case that it does not lead to a plant-wide disturbance. The main purpose of this experiment is to test if the methods will detect false causal dependencies in the generated data set.

5.2.1 Experiment 1: Pump with Loose Electricity Connection

In the first experiment the pump is set to 50 % of the maximum feeding rate and is randomly activated and deactivated to simulate a loose electricity connection. To generate the data set, the minimum time the pump is turned off is set to 20 s and the maximum time to 100 s. The valve is kept open to 50 % during the complete data acquisition. In total, a data set containing $K = 928$ samples is generated. The acquired measurements are shown in figure 5.4. Since the pump is a major process device in the system, all devices measure strong perturbations compared to the run without faults. The pressure is reduced to a minimum of nearly 5 mbar and the flow is reduced to $0\,\mathrm{m^3/min}$. In other words the loose electricity connection causes that after some seconds no water is left in the upper tank. The level sensor, positioned on the lower tank measures a minimum value of approximately 0.35 m and a maximum value of around 0.7 m. It is expected that the proposed methods detect the pump as the root cause in the resulting causal matrix. As the valve is kept on a constant value no causal dependencies should point towards or from it.

Causal matrix To calculate the causal matrix for the reconstruction of the disturbance propagation path, all methods are used on the data set and merged to the final causal matrix Q_fus. For all methods the mean causal matrix $\overline{Q}$ is set to 0.5, which results in $\beta_\mathrm{CCF} = 0.32$, $\beta_\mathrm{TE} = 0.25$, $\beta_\mathrm{GC} = 0.29$ and $\beta_\mathrm{SVM} = 0.96$ for the fitting parameters.

The results are outlined in figure 5.5. Additionally, red squares are drawn which indicate the expected causal dependencies from process knowledge. The doughnut chart and the partially directed graph indicate that the flow rate of the pump has a strong causal influence on the remaining three process variables pressure, flow and filling level. Furthermore, the methods detect for the pressure a strong causal impact on level and flow and a weak false impact on the pump. Regarding the flow, the methods detect an impact on the pressure and the filling level and a wrong weak influence on the pump. Concerning the level, a causal influence pointing to the flow and a weak influence arising on the pump and the pressure are detected.

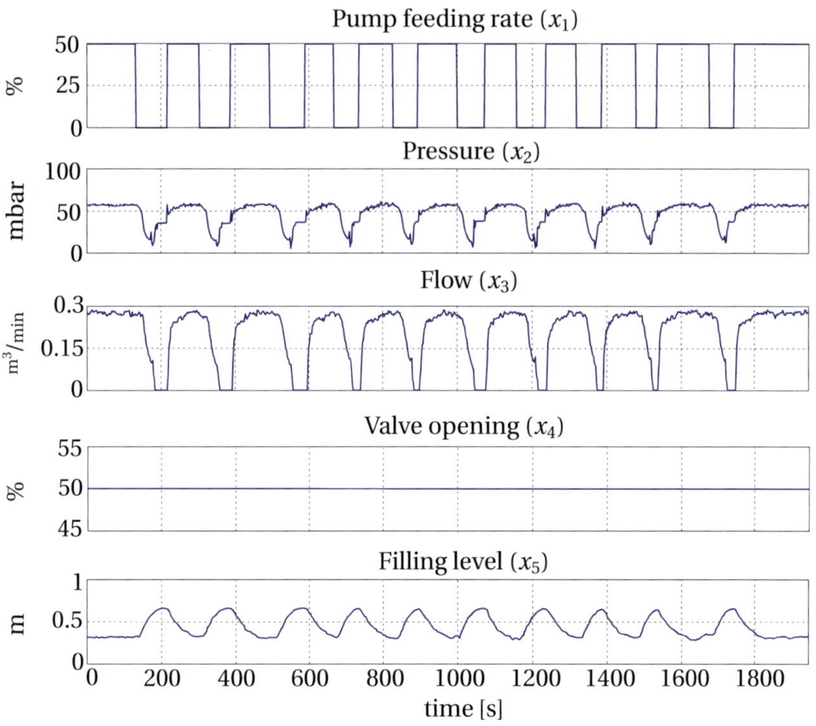

Figure (5.4) Measurement data of the laboratory plant when generating a loose electricity connection in the pump. The fault in the pump leads to disturbances in all process variables.

To compare the performance of the different methods against each other the bar chart is visualized on the right hand side. This depicts that the expected causal dependencies are all detected by several methods with a similar causal strength.

Starting with the **CCF** the results illustrate that it does not detect all expected cause-effect relationships but is also the only method that does not find false causal dependencies.

Using the **TE** three wrong causal dependencies, namely pressure → pump, flow → pump and level → flow are detected.

The **GC** detects a wrong causal dependency from level → flow but finds all other expected relationships.

All expected causal dependencies are detected using the **SVM**. Furthermore, it detects wrong causal influences pointing to the pump, to the pressure and to the flow.

To summarize, the results of this experiment indicate that when merging the methods into one resulting causal matrix the correct causal dependencies yield stronger weights and the wrongly detected causal dependencies are weighted lower and become less relevant. The transformation of Q_{fus} into the root cause priority list is shown in table 5.1. It reveals that the pump is correctly detected as being the root cause of the fault.

Rank	Process variable	RC
1	Pump feeding rate (x_1)	1.45
2	Pressure (x_2)	1.02
3	Flow (x_3)	0.75
4	Filling level (x_5)	0.43
5	Valve opening (x_4)	0

Table (5.1) Root cause priority list of the first experiment calculated from the causal matrix with the complete data set. The pump is correctly detected as being most possibly the root cause.

Receiver operating characteristic   The ROC illustrated in figure 5.6 shows the impact of the sample size when using the proposed methods. Therefore, the ROC can be interpreted as a way to investigate the time that is needed until the methods deliver a reliable outcome. To generate the functions, for each sample size the root cause priority list is calculated $N = 200$ times while using a random sample as starting point. Having $K = 50$ samples the ROC shows that using the methods gives already better results than doing a random selection of the process variables to find the root cause of a fault. While the sample size increases ($K \in \{75, 100, 200\}$) the quality of the methods when calculating the ROC increases as well, meaning that the results of the methods become more and

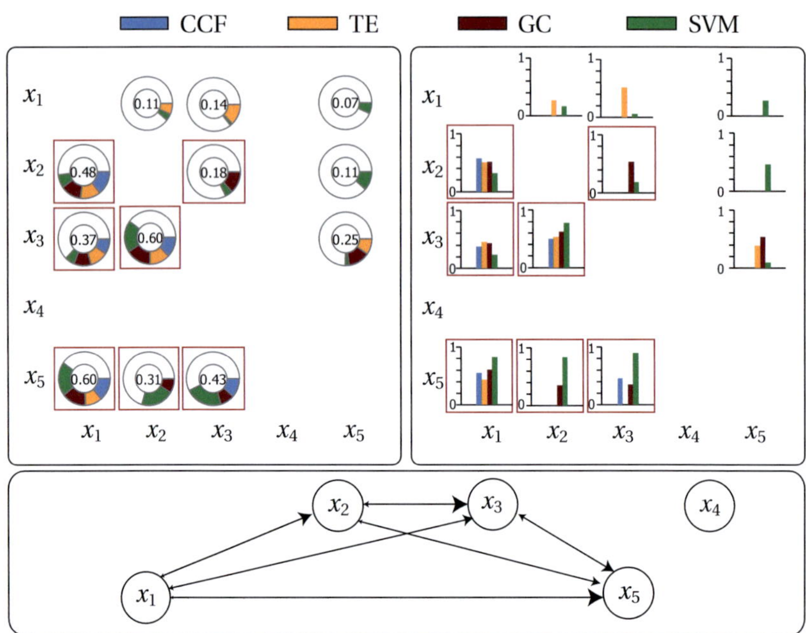

Figure (5.5) Causal matrices when having a loose electricity connection (experiment 1). The left plot shows the combined causal matrix, the plot on the right hand side compares each method in terms of a bar chart. The resulting disturbance propagation paths are visualized in the lower plot. Red squares represent the expected causal dependencies in the measurement data. (x_1 = pump feeding rate, x_2 = pressure, x_3 = flow, x_4 = valve opening, x_5 = filling level)

more reliable. Having $K = 500$ samples available, the pump is always detected as the root cause of the disturbance resulting in the theoretical maximum of the ROC. In that case the outcome mainly corresponds to the root cause priority list outlined in table 5.1 where the complete data set was used.

5.2.2 Experiment 2: Valve Air Pressure Leak

In this experiment the air pressure hose between valve and compressor is removed and reattached randomly to simulate an air pressure leak. The minimum time the connection cable is removed is set to 20 s and the maximum time is

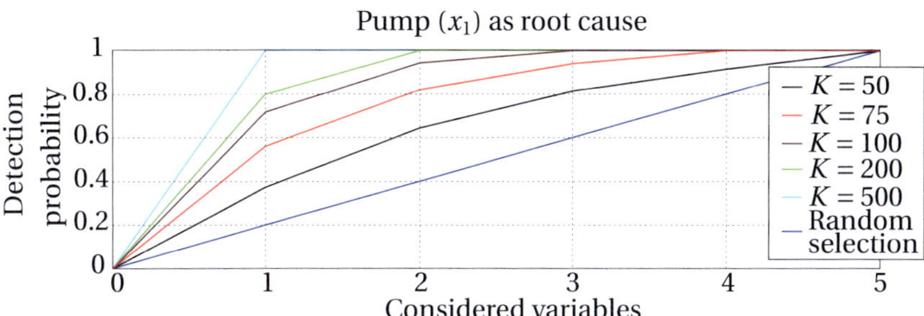

Figure (5.6) Calculated ROC for the first experiment having a loose electricity connection. The functions illustrate the impact when having different lengths of data for the root cause detection at hand.

set to 60 s. While the experiment is performed, the pump is set on 50 % of its maximum feeding rate. The resulting measurement data is illustrated in figure 5.7. In total, 800 samples are generated. Compared to the first experiment, the disturbances in the measurements of the process variables show a much more random behavior.

When the valve closes, the water is blocked from flowing from the upper to the lower tank. Since the pump runs with a constant feeding rate, the hydraulic pressure increases as more water enters the upper tank. When the valve reopens, the water stored in the upper tank can flow to the lower tank and the process goes back into the stationary phase without a fault. This behavior becomes obvious in further detail by analyzing the measurements of the process variables. After the valve closes, the flow reduces to $0\,\mathrm{m^3/min}$ while the level meter measures a continuous reduction of the water in the lower tank. The hydraulic pressure increases until the pump stops delivering water from the lower to the upper tank. At the moment the pump stops, the level sensor measures about 0.15 m. The expected disturbance propagation path should detect the valve as root cause as it has an effect on all process variables. Compared to flow and pressure meter the level sensor should detect the closing of the valve later. The reasons are that the tube between valve and lower tank needs to be emptied before the disturbance can be measured and therefore has an effect like a dead time and that the lower tank acts as a low-pass filter.

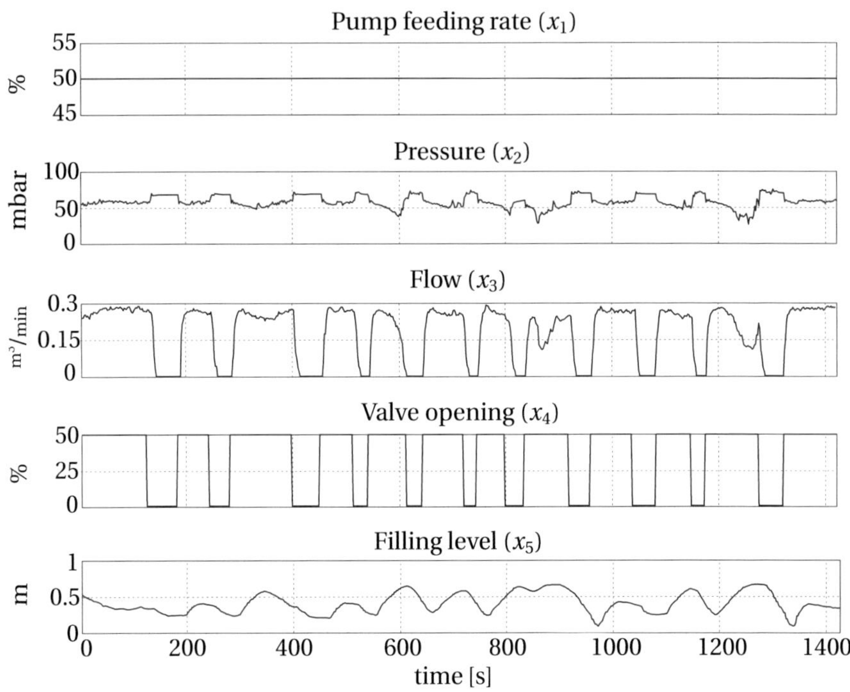

Figure (5.7) Data of the laboratory plant when simulation a faulty valve (experiment 2). Compared to the first experiment the fault leads to much more random disturbances.

Causal matrix   The generated causal matrix Q_{fus} of the experiment using all suggested methods is visualized in figure 5.8. Like for the first experiment, for all methods the mean causal matrix $\overline{Q}$ is set to 0.5. This results in $\beta_{\mathrm{CCF}} = 0.33$, $\beta_{\mathrm{TE}} = 0.26$, $\beta_{\mathrm{GC}} = 0.31$ and $\beta_{\mathrm{SVM}} = 0.82$ for the fitting parameters.

The results show a strong causal impact from the valve pointing towards the remaining process variables. Furthermore, all process variables show a strong causal dependency pointing towards the filling level. The filling level again is considered as having a weak cause-effect relationships pointing towards the other process variables. As in the first experiment the pressure is evaluated to have a stronger impact on the flow than vice versa.

The bar chart, used to compare the performance of the different methods against each other, is visualized on the right hand side in figure 5.8. All known causal dependencies, marked as red squares, were detected by more than one method.

Regarding the methods this shows that the **CCF** detects all three causal dependencies pointing from the valve to the other process variables and in addition finds the two causal dependencies pressure $\rightarrow$ level and pressure $\rightarrow$ flow. Furthermore, a false causal dependency level $\rightarrow$ flow is indicated by the CCF.

Except for the cause-effect relationship pressure $\rightarrow$ level, the **TE** finds only intercausal dependencies in the data meaning that the variables all have a causal influence onto each other.

The **GC** detects all expected causal dependencies except the dependency in which flow and valve opening point towards the filling level. In addition, it detects a wrong weak causal dependency pointing from level $\rightarrow$ valve.

Except for the cause-effect relationship flow $\rightarrow$ pressure, the **SVM** finds all expected dependencies. False causal dependencies are the level pointing towards the pressure and the flow.

The outcome shows again that when merging the methods into one resulting causal matrix the correct causal dependencies obtain in a stronger weighting meaning that the wrongly detected causal dependencies from some methods become less relevant.

Finally, Q_{fus} is transformed into the root cause priority list outlined in table 5.2. The valve is correctly detected as the root cause of the fault.

Rank	Process variable	RC
1	Valve opening (x_4)	1.61
2	Pressure (x_2)	1.04
3	Flow (x_3)	0.69
4	Filling level (x_5)	0.61
5	Pump feeding rate (x_1)	0

Table (5.2) Root cause priority list for the second experiment calculated from the causal matrix using the whole data set. The valve is correctly found as being most probably the root cause.

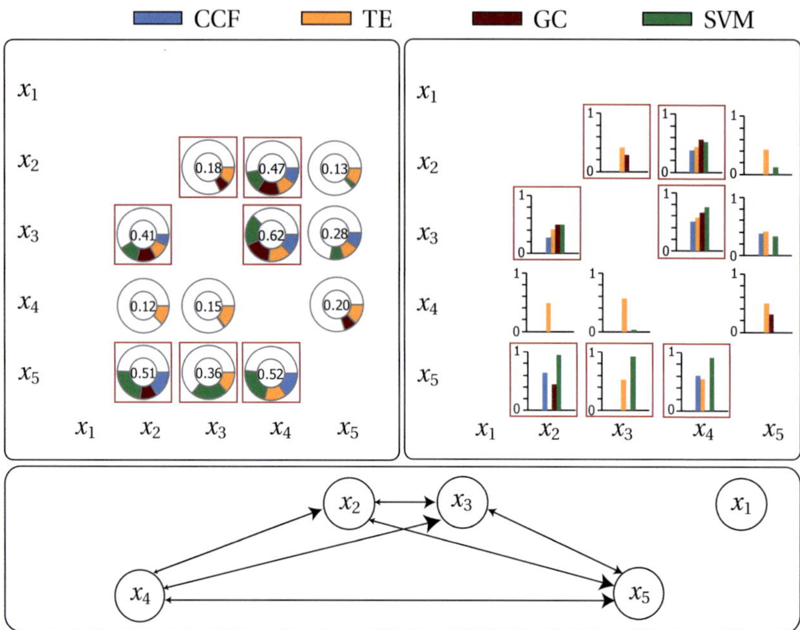

Figure (5.8) Causal matrices when simulating an air pressure leak (experiment 2). The left plot shows the combined causal matrix, the plot on the right hand side each method in terms of a bar chart. The resulting disturbance propagation paths are illustrated in the lower plot. Red squares represent the expected causal dependencies for this experiment. (x_1 = pump feeding rate, x_2 = pressure, x_3 = flow, x_4 = valve opening, x_5 = filling level)

Receiver operating characteristic   Like in the first experiment, the ROC is used to show the impact of the sample size on the detection of the valve as root cause. The resulting ROC for the sample sizes $K \in \{50, 75, 100, 200\}$ is given in figure 5.9. The result can be interpreted as the time that is needed until the methods deliver reliable results when a fault occurs in the plant. For each sample size the root cause priority list is calculated $N = 200$ times while using a random sample as starting point. Having $K = 50$ samples at hand the ROC shows already good results as a selection according to the function is significantly better than a random selection of the process variables. While the sample size increases ($K \in \{75, 100\}$) the quality of the methods when calculating the ROC increases as well, resulting for $K = 200$ samples in the theoretical maximum of the ROC. For this sample size the valve is detected in all data sets as the root cause which corresponds to the root cause priority list outlined in table 5.2. Compared to the first experiment this illustrates that less samples are needed to detect the fault reliably.

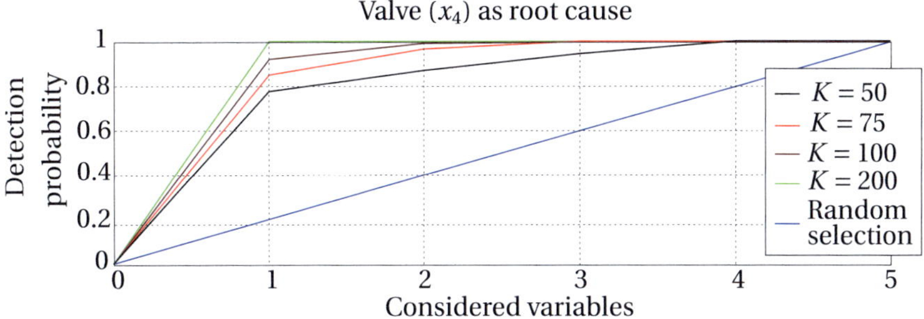

Figure (5.9) ROC for the second experiment when having as fault an air pressure leak. The functions show the impact when having different lengths of data available for analysis. For $K = 200$ samples this results in the theoretical maximum of the ROC.

5.2.3 Experiment 3: Fault in Pump and Consecutive Fault in the Valve

In this experiment it is assumed that pump has a fault and the valve results in a consequential fault. The main aim of this experiment is to investigate if the methods can still detect the pump as root cause even if the valve is acting as a second actuator on the system. To generate this kind of fault the flow rate of the pump is reduced to 30 % and set back to its normal set-point, while the valve closes down to 50 % and reopens to 70 % each time with a 15 s delay regarding the pump. This procedure is repeated once. Compared to the first two experiments two causes are now acting on the system while the pump should be detected as root cause and the valve ranked in second place in the root cause priority list. The measurement data acquired from the experiment is given in figure 5.10 and can be explained the following way. Since the flow rate of the pump is reduced, less water is delivered into the upper tank, while a partial closing of the valve leads to a lower outflow. This results in fluctuations which are detected by all three sensors. The flow meter measures a reduced water flow when the pump reduces its flow rate and the valve closes. The pressure sensor measures a slight drift but has less fluctuation compared to the first two experiments. The filling level in the lower tank slightly increases as the pump delivers less water to the upper tank. At the moment, the pump and valve are set back to their original values, this causes that the level in the lower tank decreases again.

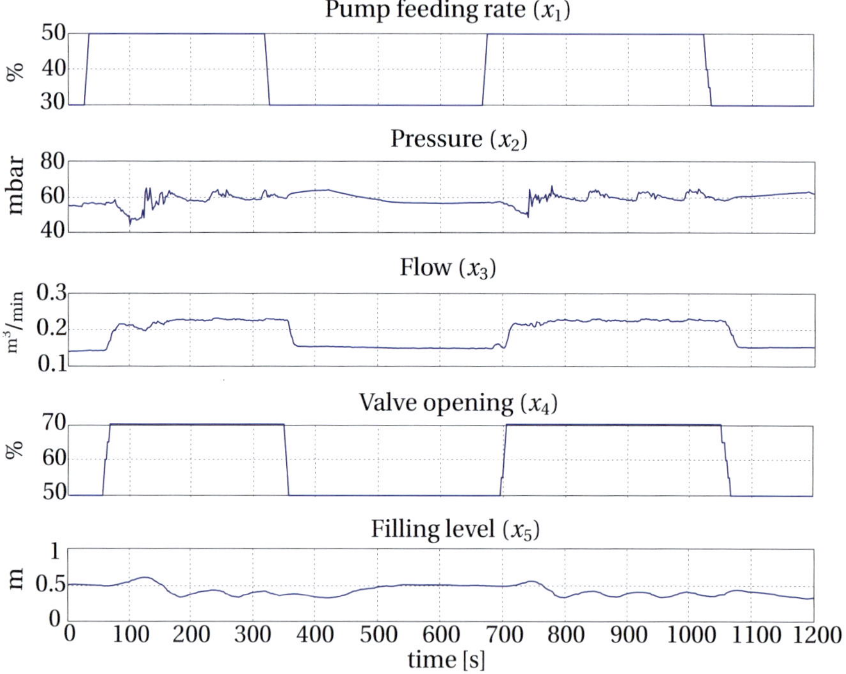

Figure (5.10) Data of the laboratory plant when generating a combined fault in the pump and the valve (experiment 3). The valve follows the behavior of the pump with a 15 s delay. Therefore, the pump should be detected as root cause.

Causal matrix   The resulting causal matrices are outlined in figure 5.11. Like in the other experiments, for all methods the mean causal matrix $\overline{Q}$ is set to 0.5. For the fitting parameters this results in $\beta_{CCF} = 0.24$, $\beta_{TE} = 0.48$, $\beta_{GC} = 0.56$ and $\beta_{SVM} = 0.78$.

Compared to the first two experiments the causal dependencies are less clear. The combined causal matrix shows that the pump feeding rate has a strong causal impact on the valve, which is reasonable as the valve follows the movement of the pump with 15 s delay. Another strong causal impact from the pump is detected pointing towards the pressure and a weak one towards the flow. Like in the other experiments, the pressure has a stronger causal impact on the flow than vice versa. Furthermore, in terms of the pressure two weak false causal dependencies are detected pointing towards pump and valve. Regarding the valve, the combined causal matrix indicates a strong causal impact on the flow and weak causal dependencies for pressure and level sensor. Finally, the filling level can be interpreted as an output variable, as the methods detect only weak cause-effect relationships pointing towards the pressure and the valve.

The bar chart used to compare the methods is outlined on the right hand side in figure 5.11. Since there are now two process variables acting as a fault on the plant, this leads to more detected intercausal dependencies compared to the first two experiments. The expected dependencies from process knowledge are marked as red squares in the causal matrix.

Passing the methods one-by-one, the **CCF** detects two false causal dependencies pointing from pressure to valve and from level to pump. Furthermore, it detects six times the causal dependencies expected from process knowledge. Therefore, the outcome shows that especially the pump is detected as input and the level meter as output variable.

The **TE** yields one false causal dependency pointing from the pressure to the pump. The other five found cause-effect relationships are all expected.

Regarding **GC**, four different causal dependencies are detected which are all expected.

Finally, the **SVM** finds three causal dependencies, among which the causal dependency pointing from level to pressure is wrong.

The transferred causal matrix in terms of the root cause priority list is given in table 5.3. It reveals, that as expected the pump is ranked at position one,

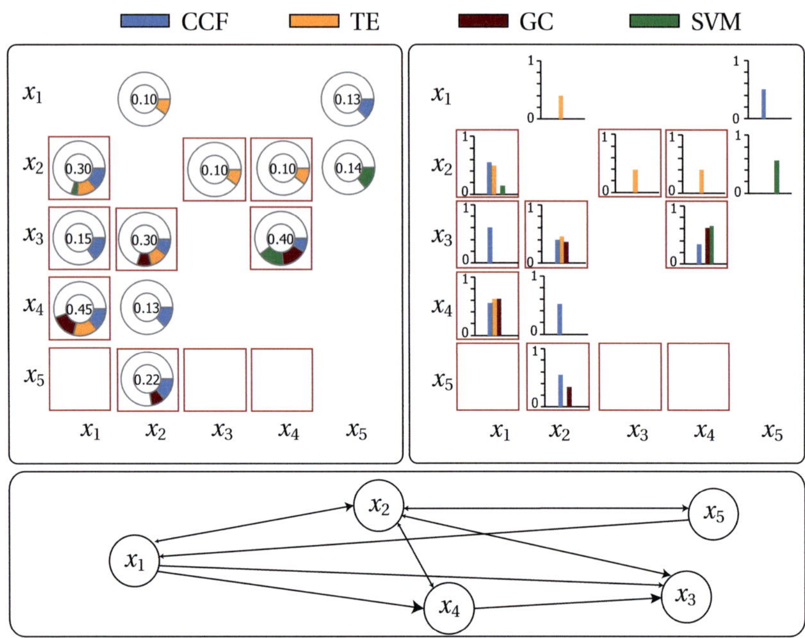

Figure (5.11) Causal matrices for a combined fault in the pump and the valve (experiment 3). The left plot illustrates the combined causal matrix, the plot on the right hand side compares each method in terms of a bar chart. The resulting disturbance propagation paths are outlined in the lower plot. Red squares represent the expected causal dependencies. (x_1 = pump feeding rate, x_2 = pressure, x_3 = flow, x_4 = valve opening, x_5 = filling level)

meaning that the correct root cause has been found. Still, comparing this list with the resulting priority list from the first experiment (table 5.1) indicates that the pump being the root cause of the fault is not as obvious as in the former experiment.

The valve, being the second variable acting on the system is ranked on position three. The reason is, that regarding the pressure two false causal dependencies have been found which give this sensor a higher weight compared to the valve. In terms of the expected fault propagation path, this is not correct.

Rank	Process variable	RC
1	Pump feeding rate (x_1)	0.90
2	Pressure (x_2)	0.75
3	Valve opening (x_4)	0.50
4	Filling level (x_5)	0.27
5	Flow (x_3)	0.10

Table (5.3) Root cause priority list for the third experiment calculated from the causal matrix using the complete acquired measurement data. The pump is correctly detected of being most possibly the root cause.

Receiver operating characteristic   The ROC is illustrated in figure 5.12 and shows the impact of the sample size on the detection of the pump as root cause. For each sample size the root cause priority list is calculated $N = 200$ times while using a random sample as starting point. Compared to the first two experiments more samples are needed for the correct detection of the pump as root cause. The functions generated from $K = 75$ and 100 samples do not significantly differ from a random selection of the process variables. For this experiment at least $K = 150$ samples are needed so that the function starts to differ from a random selection. Increasing the sample size further ($K \in \{200, 500\}$) shows that the function of the ROC leads to much better results compared to a random selection. Still the theoretical optimum cannot be reached even when having $K = 500$ samples available.

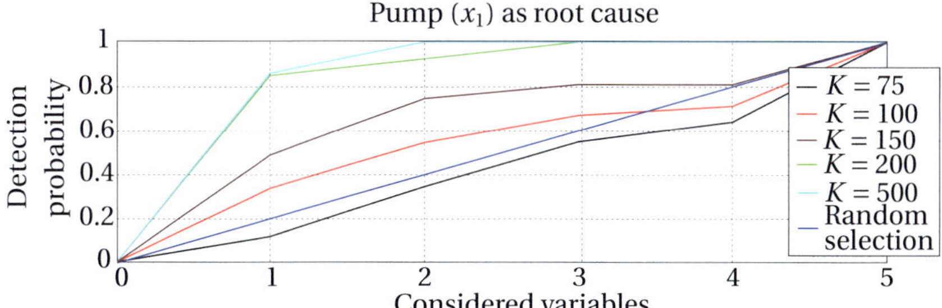

Figure (5.12) ROC for the third experiment when generating a fault in the pump followed by a fault in the valve. The functions illustrate the impact for the detection of the pump as root cause when having different lengths of disturbed measurement data for analysis.

5.2.4 Experiment 4: Oscillating Pump with Loose Electricity Connection

In this experiment the pump follows an oscillating waveform while applying as a fault two times a loose electricity connection. Due to the fact that the waveform is cyclic, no causal information can be drawn from it and all cause-effect relationships need to be detected through the fault in the pump. One sinusoidal oscillation takes around 700 s. After the engine drop-out, the pump retakes again its oscillating waveform. During the whole run the valve is kept 50 % open. The measurement data is displayed in figure 5.13. Each of the two drop-outs has a measurable impact on the other process variables. In detail the pressure meter stays on a value of approximately 60 mbar while the pump runs in its oscillating mode. The moment the fault occurs, the hydraulic pressure shows a sharp peak before settling back to 60 mbar. The flow shows a similar behavior as the acquired measurements with sharp peaks. Furthermore, the flow meter slightly changes its set-point after the first fault. In detail the flow changes from around $0.17\,\mathrm{m^3/min}$ down to $0.15\,\mathrm{m^3/min}$. Like the other process variables the level meter measures a peak as the fault in the pump occurs. In that case the level increases from $0.25\,\mathrm{m}$ to nearly $0.4\,\mathrm{m}$ the moment the fault occurs.

Causal matrix   The resulting causal matrices are shown in figure 5.14. The expected causal dependencies are all marked as red squares. The mean causal matrix $\overline{Q}$ is set to 0.5 for all methods, which results for the fitting parameters in $\beta_{\mathrm{CCF}} = 0.34$, $\beta_{\mathrm{TE}} = 0.47$, $\beta_{\mathrm{GC}} = 0.22$ and $\beta_{\mathrm{SVM}} = 0.21$.
Compared to the first experiment, which has the pump as root cause as well, the merged causal matrix should give the same disturbance propagation path. Still, the results show that due to the oscillations the found cause-effect relationships are less obvious. Regarding the pump, causal dependencies pointing towards pressure and flow are detected, but the causal dependency towards the filling level of the lower tank are not found. In terms of the pressure, the methods detect the two expected causal dependencies pointing towards filling level and flow. The flow shows the expected causal dependencies that point towards the pressure sensor and the filling level. Furthermore, a wrong causal dependency

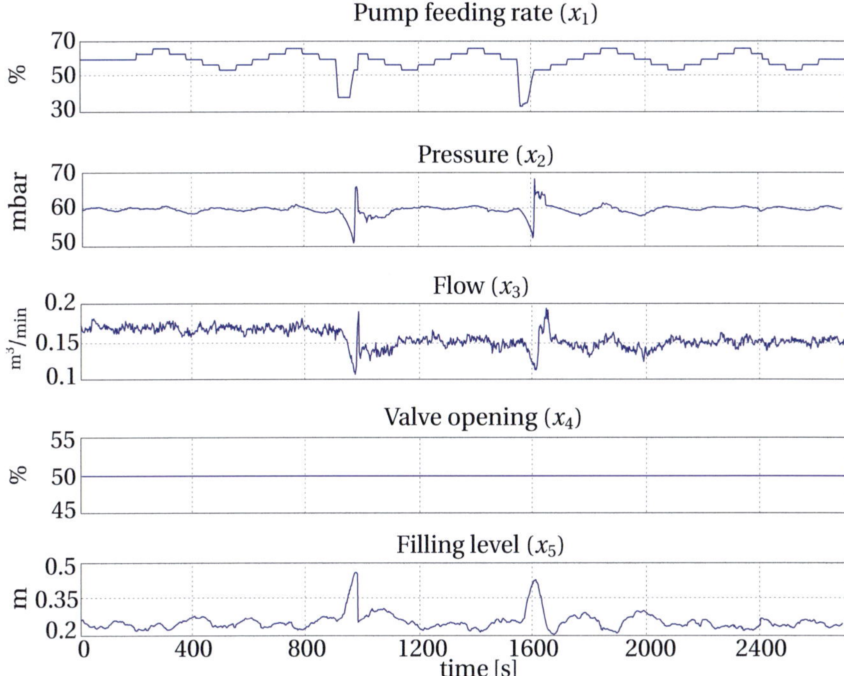

Figure (5.13) Data of the laboratory plant when having a fault in the flow rate of the pump while the pump performs a sinusoidal waveform (experiment 4).

pointing from flow to pump has been found. Finally, the proposed methods find causal dependencies pointing from the filling level towards pump, flow and pressure. These are all not correct as the filling level is the variable of the process that notices the disturbance last.

Passing the methods one-by-one, the **CCF** detects four of the expected causal dependencies. Furthermore, a false dependency is found pointing from flow to pump.

The **TE** finds an intercausal dependency between the pressure and the flow and a false causal dependency pointing from the level to the pump.

Similar to the CCF the **GC** detects the same four correct causal dependencies.

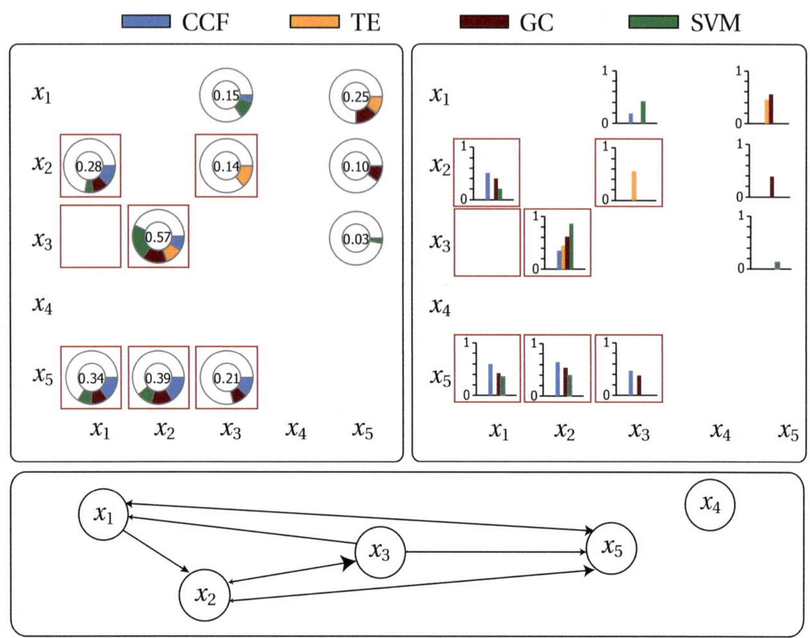

Figure (5.14) Causal matrices when having an oscillating pump with a loose electricity connection as root cause (experiment 4). The left plot shows the combined causal matrix, the plot on the right hand side shows each method in terms of a bar chart. The resulting disturbance propagation paths are given in the lower plot. (x_1 = pump feeding rate, x_2 = pressure, x_3 = flow, x_4 = valve opening, x_5 = filling level)

Still, it detects false causal dependencies pointing from the filling level towards pump and pressure.

Finally the **SVM** gives similar results as CCF and GC. It detects four times correct causal dependencies and two times a wrong cause-effect relationship which is in detail level → flow and flow → pump.

Table 5.4 outlines the root cause priority list generated from the causal matrix. The pressure is detected as root cause which is mainly due to the fact that a strong causal dependency is detected by all methods pointing from pressure to level meter. The pump, which is the real root cause of the fault is ranked on position two. Compared to the first experiment this illustrates the detrimental influence of oscillations when performing a root cause analysis.

Rank	Process variable	RC
1	Pressure (x_2)	0.96
2	Pump feeding rate (x_1)	0.62
3	Flow (x_3)	0.5
4	Filling level (x_5)	0.38
5	Valve opening (x_4)	0

Table (5.4) Root cause priority list of the fourth experiment calculated from the causal matrix using the whole data set. The pressure is most probable being the root cause, the pump which is the real root cause of the fault is ranked second.

Receiver operating characteristic   The ROC, illustrated in figure 5.15, shows the impact of the sample size for the detection of the pump as root cause. For each sample size the root cause priority list is calculated $N = 200$ times while using a random sample as starting point from the available data set. The starting point of the first sample is set to 400 to assure that the selected samples contain at least once the fault. The pump is ranked on position two when analyzing the whole data set which means that the theoretical maximum cannot be achieved. The result shows that the functions for the low sample sizes $K \in \{100, 150\}$ do not significantly differ from a random selection. Starting with $K = 200$ samples the ROC yields a relevant difference compared to a random selection of the process variables. When the sample size is increased further to $K = 500$ samples, the function of the ROC delivers the best results as with a probability of approximately 60 % the pump is detected as root cause when selecting the first ranked variable.

5.2.5 Experiment 5: Tube Clogging

In this experiment a stop cock, which is situated between the flow sensor and the valve, is once partially closed and reopened to have for some time a water flow reduction in the plant. The pump is kept on 50 % of its maximum feeding rate and the valve is kept 50 % open. In other words, no fault occurs in an active plant component and the measured disturbances are generated from an

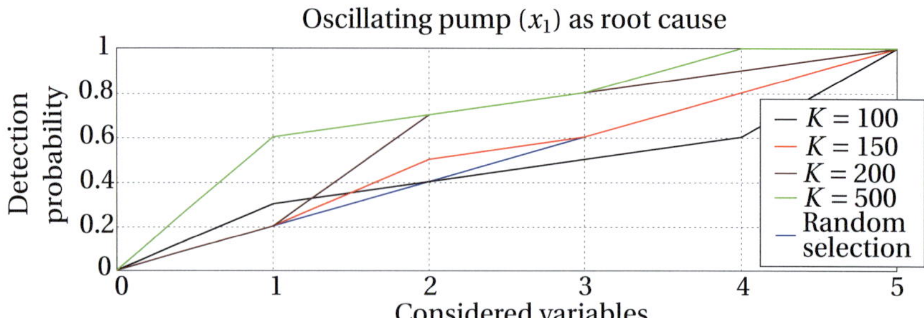

Figure (5.15) ROC for the fourth experiment when having a loose electricity connection in an oscillating pump as root cause. The functions show the impact of different lengths of data for analysis.

external fault. The measured data is illustrated in figure 5.16. In total, $K = 450$ samples are available for analysis.

The moment the stop cock is partially closed, less water flows from the upper tank to the lower tank. Since the pump delivers water with a constant flow rate into the upper tank, the pressure sensor measures an increase of the hydraulic pressure and the flow meter a reduced water flow. The lower tank is partially emptied and the plant changes to another stationary phase. When the stop cock is reopened the plant switches back to the stationary phase it had before closing the stop cock.

The behavior of the disturbance is similar to the one in experiment 3, in which an air pressure leak in the valve was simulated. But in contrast, in this case there is no measurement data available from the valve which acts as the root cause of the fault. In addition, the disturbance is affecting the process only in terms of two step changes and not like the air pressure leak in steps with random length. As the tube clogging can be seen as an external fault and since the stop cock is situated directly behind the flow meter, the pressure sensor and the flow meter should detect this disturbance first.

Causal matrix   The resulting causal matrix is outlined in figure 5.17. The red squares in the doughnut and the bar chart mark the expected causal dependencies from process knowledge. Again, the mean causal matrix $\overline{Q}$ is set to 0.5 and

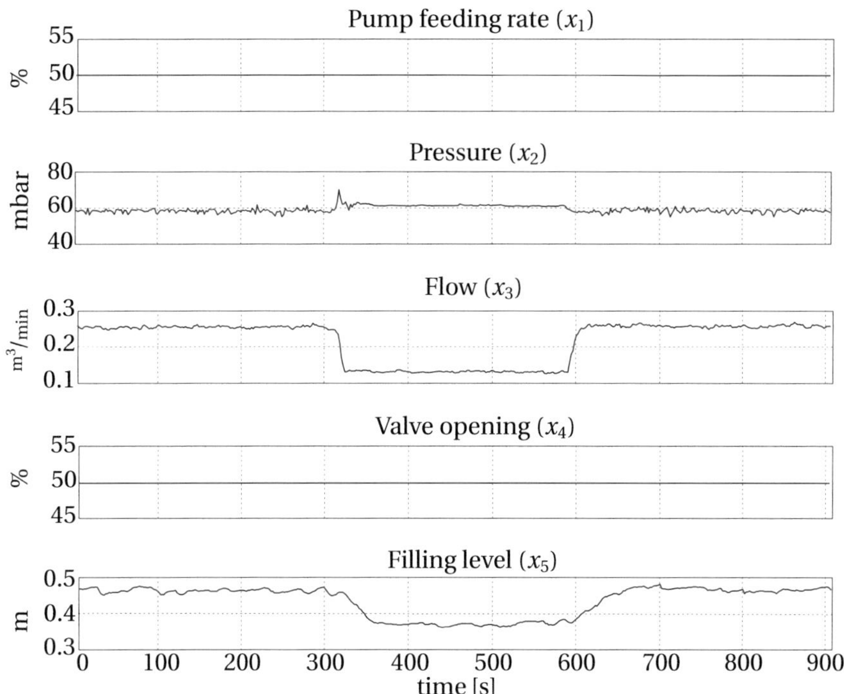

Figure (5.16) Data of the laboratory plant when simulating tube clogging (experiment 5). In detail, a stop cock which is situated between the flow meter and the valve is once partially closed and reopened to simulate a reduced water flow in the plant.

the fitting parameters for the different methods result in $\beta_{CCF} = 0.23$, $\beta_{TE} = 0.21$, $\beta_{GC} = 0.29$ and $\beta_{SVM} = 1.02$.

Pump and valve are kept on a constant value during the experiment, which means that there are no causal dependencies to be detected from their part. Regarding the pressure, causal dependencies pointing towards the flow and the filling level are found. For the flow, causal impacts are found pointing towards the pressure and the filling level. Finally, the developed methods detect a false causal dependency pointing from the filling level to the flow.

The bar chart on the right hand side shows the outcome of the different meth-

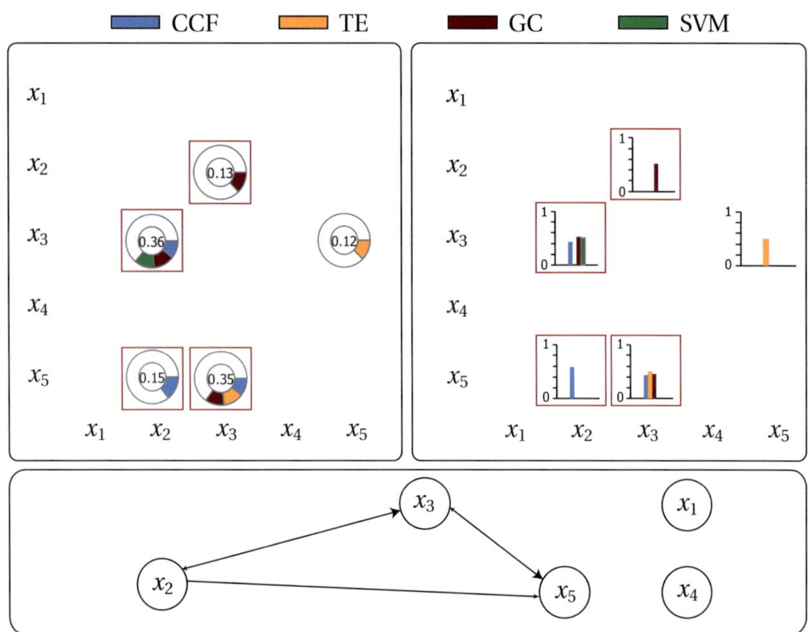

Figure (5.17) Causal matrices when simulating tube clogging (experiment 5). The left plot represents the combined causal matrix, the plot on the right hand side compares each method in terms of a bar chart. The found disturbance propagation paths are given in the lower plot. Red squares represent the expected causal dependencies for this experiment. (x_1 = pump feeding rate, x_2 = pressure, x_3 = flow, x_4 = valve opening, x_5 = filling level)

ods. In case of the **CCF** three causal dependencies are detected which are all expected ones and it misses only the dependency pointing from flow to pressure.

The **TE** is the only method that detects the wrong cause-effect relationship from filling level to flow but finds also the correct direction flow $\rightarrow$ level.

Similar to the CCF the **GC** detects three causal dependencies, which are all correct. It misses only the dependency pressure $\rightarrow$ level.

Finally, the **SVM** finds the cause-effect relationship pressure $\rightarrow$ flow.

The generated root cause priority list is given in table 5.5. This reveals that the pressure has been detected as being the root cause, as it has a slightly stronger weight compared to the flow. The filling level shows the lowest weighting, as valve and pump were set on constant values during the experiment. This outcome is reasonable, as the stop cock is situated in the tube behind the flow meter and in front of the valve. Therefore, the pressure sensor as well as the flow meter should react at the same time on the disturbance.

Since the root cause comes from an external disturbance, this experiment can be interpreted as a test which sensor detects the disturbance first but does not represent the actual cause of the fault. Therefore, no receiver operating characteristic is calculated.

Rank	Process variable	RC
1	Pressure (x_1)	0.51
2	Flow (x_2)	0.48
3	Filling level (x_4)	0.12
4	Valve opening (x_3)	0
5	Pump feeding rate (x_5)	0

Table (5.5) Root cause priority list of the fifth experiment calculated from the causal matrix when simulating tube clogging. The pressure sensor is the first process variable detecting the fault as it results in a slightly higher value than the flow. It was expected that both process variables result in similar values.

5.2.6 Experiment 6: Faulty Level Sensor

In this experiment the level sensor is moved up and down to simulate an error in the measurement device. The pump is set to 50 % of its maximal feeding rate during the whole run and the valve is kept open 50 %. The measurement data is displayed in figure 5.18. Since the level meter measures the height of the water in the lower tank and as there is no feedback to an actuating variable, the disturbance does not propagate through the plant. For that reason, the other process variables yield similar measurements as illustrated in figure 5.3 where the process runs without faults. Since this fault does not lead to a plant-wide disturbance, it is expected that the methods do not detect any causal dependency in the data.

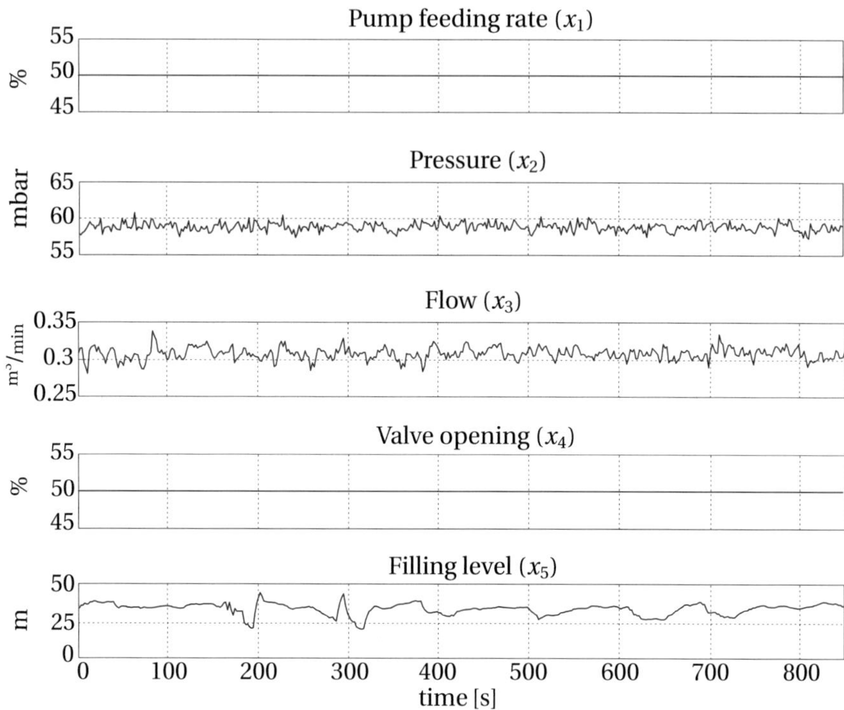

Figure (5.18) Data of the laboratory plant when having a faulty level sensor (experiment 6). Since a faulty level sensor does not lead to a plant-wide disturbance, no causal dependencies are to be found using the proposed methods.

Causal matrix None of the methods detected any causal relationship between the process variables when using the measurement data from this experiment. This means, that for each method all significance tests for the causal dependencies failed. Therefore, the combined causal matrix Q_{fus} leads to a matrix which contains only zeros, which confirms the expected result from the fault. This result shows that the application of significance tests is reasonable, as otherwise all methods would return values with low causal strengths. These values could then be falsely interpreted as the disturbance propagation path of the fault which is not existing in this experiment.

5.3 Summary

In this chapter the developed methods were tested by generating faults on a laboratory plant. The plant is characterized by five different process variables and is used to pump water around in cycles. At all, six distinct experiments were performed to investigate the behavior of the developed algorithms. In summary, the results showed that, except for one case, the process variable being the root cause of the disturbance could be correctly detected.
Due to the combination of the methods, more robust results are achieved than using only one single method at a time. The reason is, that causal dependencies wrongly detected by one method, result in low weights and therefore have only a small impact on the calculated root cause priority list. Furthermore, the results showed that the best working method is strongly dependent on the disturbances within the measurement data. In other words, none of the suggested methods could outperform the others when using data coming from this plant. This again clarifies the advantage of the combination of the methods to one resulting causal matrix.
A receiver operating characteristic was used to check the number of samples that is needed to reliably detect the faulty process variable. In that case, the results revealed that the sample size needed for the root cause detection is strongly dependent on the complexity of the occurred fault. If there is only one faulty process variable acting on the process and in addition the plant is in a stationary phase, less data is needed in comparison to having oscillations or more than one faulty variable acting on the plant.

6

Industrial Glass Forming Process

In this chapter the proposed methods are used to reconstruct the propagation path of a known disturbance which occurs during glass rod production in an industrial process. The chapter starts with a detailed explanation of the process characteristics and the expected disturbance propagation path on the basis of process knowledge. Thereafter, the methods are tested on two exemplary production runs containing disturbances. The chapter closes by summarizing the obtained results.

6.1 Process Characteristics

The main objective of this process is to pull thin glass rods out of thick glass cylinders, which are manufactured in a preprocessing step. This rod is used in a subsequent step for the fabrication of glass fibers and therefore its production needs to fulfill high quality standards. Figure 6.1 sketches on the left hand side the plant as well as the position of the used process equipment. The process is controlled by several process variables which are of different importance depending on the process phase of the production. Their characteristics can be explained the following way. When the cylinder enters the oven, initially the temperature ϑ_c (c = cylinder) above the oven is measured. The other process variables used for analysis are all measured below the oven. As the hot beam leaves the oven on the downside, the glass temperature ϑ_b (b = beam) and the upper diameter D_u (u = up) are measured. The pulling speed v_s is used as an actuating variable to keep D_u on its set-point. Finally, the lower diameter D_l (l = low) of the solidified glass is measured in some further distance below the measurement device acquiring D_u. Table 6.1 summarizes the process variables. Further information about the process with a special focus on the underlying control concept can be found e. g. in [SK97].

Variable	Description
ϑ_c	Cylinder temperature
D_u	Beam upper diameter
D_l	Beam lower diameter
ϑ_b	Beam temperature
v_s	Pulling speed

Table (6.1) Process variables for the calculation of the disturbance propagation path in the investigated industrial glass forming process.

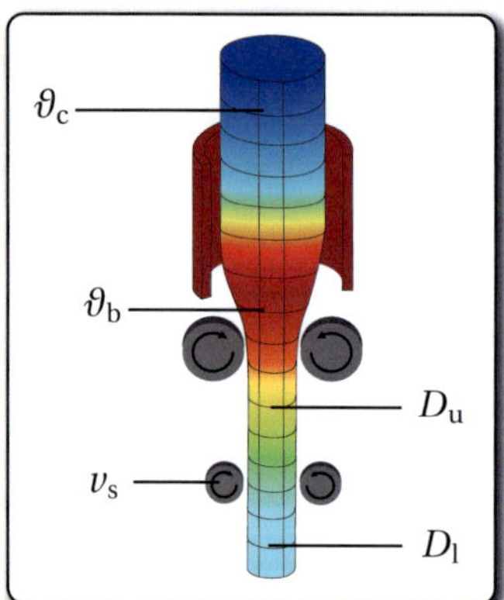

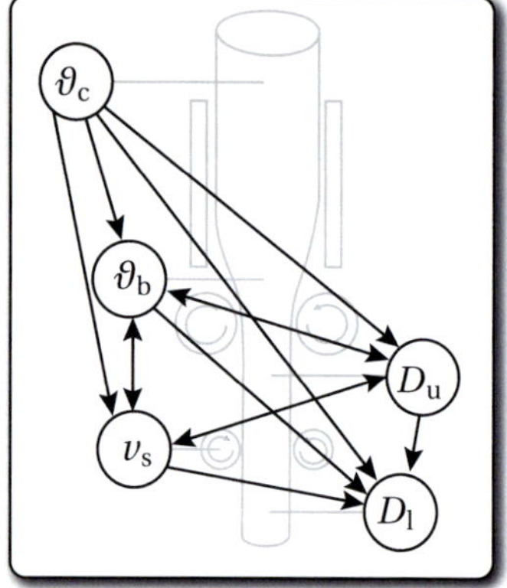

Figure (6.1) Industrial glass forming plant used for the calculation of the disturbance propagation path resulting from a weld distortion in the glass cylinder. The left plot sketches the plant and the used process variables, the right plot illustrates the disturbance propagation path originating from process knowledge in terms of a partially directed graph.

Disturbance propagation path   To attain an efficient production in some cases two cylinders are welded together and used for the production of the glass rod. As the cylinder weld shows a different thermal behavior compared to the rest of the glass, this leads to a distortion which propagates through the process. The characteristics of the disturbance propagation path and its impact on the different process variables is understood in general and therefore the process is suited to test the developed methods on this industrial plant. In detail it is checked if the proposed methods can be used to reconstruct the disturbance propagation path when the weld distortion propagates through the different process variables.

The expected propagation path from process knowledge is illustrated in figure 6.1 on the right hand side in terms of a directed graph. The explanation of the dependencies is done by taking into account the data of the two investigated production runs displayed in figure 6.2 and 6.4. These productions illustrate the impact of the weld disturbance connecting the two cylinders as all process variables show distortions while it propagates through the process. The first variable which detects the disturbance is the cylinder temperature ϑ_c as it is measured above the oven. It is the first variable that starts fluctuating which means that it should result as being the root cause variable and have a causal impact on all other process variables. The moment the weld approaches the lower end of the oven, the temperature of the glass ϑ_b and the measurement device of the upper diameter D_u measure the distortions coming from the weld. Furthermore, as the underlying control system uses the upper diameter as controlled value, the pulling speed v_s starts acting against the disturbances to stabilize D_u. This is the reason why all three variables measure the disturbance closely together in time and it is expected that the methods detect an intercausal connectivity among ϑ_b, D_u and v_s. The lower diameter D_l is used to measure the diameter of the beam after the glass has already partially cooled down, meaning that it is the last variable detecting the disturbance. This becomes obvious by looking at figure 6.2 and 6.4. The data of the productions illustrates that D_l follows the behavior of D_u with some delay in time. Therefore, it is expected that ϑ_c, v_s, D_u and ϑ_b all have a causal dependency pointing towards D_l.

6.2 Propagation Analysis of Weld Disturbance

The proposed methods in chapter 3 are used to reconstruct the disturbance propagation path coming from the weld disturbance explained in the section above. As the weld passes the plant, the process variables detect the disturbance at different points in time and therefore it is possible to calculate the cause-effect relationships. From this, it is possible to reconstruct the disturbance propagation path.

Two productions were used to test the methods. In the first run, the weld only leads to small distortions in the data. This has the advantage that the measured disturbances are more distinct in time compared to the second run. In the second run, the distortions are much stronger, meaning that the differences are less distinct in time and it is expected that the methods will face more problems to reconstruct the disturbance propagation path.

6.2.1 Analysis of the First Production

The first measurement data set used for the reconstruction of the disturbance propagation path is displayed in figure 6.2. All proposed methods are used to analyze the data set and to calculate the causal matrices. The resulting combined causal matrix is illustrated in figure 6.3. In the doughnut and the bar chart red squares are sketched which represent the causal dependencies coming from the process knowledge displayed in figure 6.1. For the reconstruction of the disturbance propagation path, all methods are used and merged to the final causal matrix Q_{fus}. The mean causal matrix $\overline{Q}$ is set to 0.5 for all methods, which results in $\beta_{\text{CCF}} = 0.59$, $\beta_{\text{TE}} = 0.24$, $\beta_{\text{GC}} = 0.2$ and $\beta_{\text{SVM}} = 0.68$ for the fitting parameters.

Evaluating the outcome of the first production shows that most of the expected causal dependencies for the reconstruction of the disturbance propagation path could be found. The bar chart outlined on the right hand side of figure 6.3 is used to analyze and compare the methods in further detail.

Starting with the **CCF** the results show that it correctly detects ϑ_{c} as input and D_{l} as output variables. Furthermore, it does not detect any wrong causal dependency.

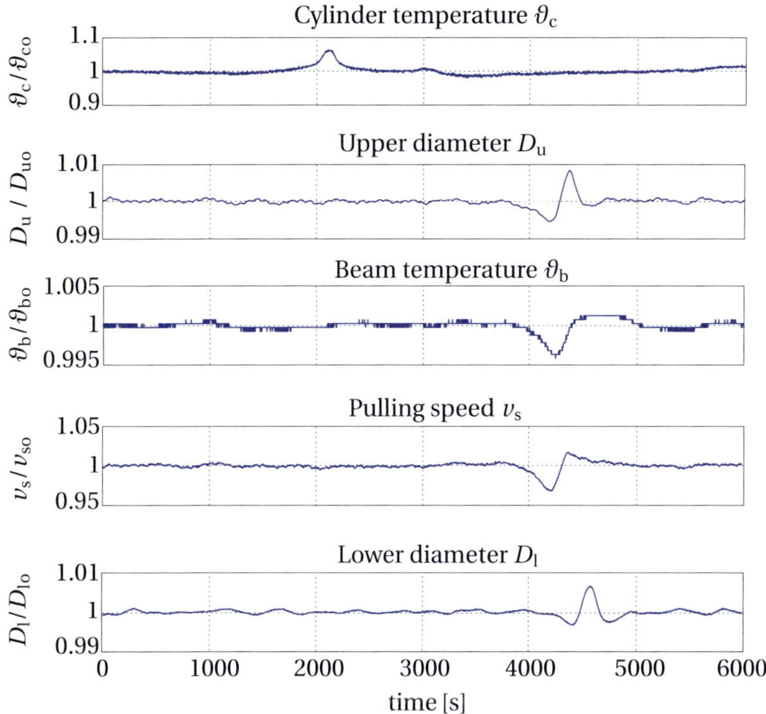

Figure (6.2) Data of the industrial glass forming process containing a weld disturbance. This data set is used as the first test for the reconstruction of the disturbance propagation path.

The **TE** does not give good results on the data set as except the not existing causal dependencies $D_u \rightarrow \vartheta_c$ and $D_l \rightarrow \vartheta_c$ the method detects intercausal relationships among all other process variables. A possible explanation is that the transfer entropy faces problems with narrow-banded disturbances (as shown in section 3.5.2) which is the case for the weld disturbance.

The **GC** detects the two causal dependencies $D_u \rightarrow D_l$ and $v_s \rightarrow D_l$. Both dependencies were expected from the process knowledge. Causal dependencies containing the cylinder temperature ϑ_c could not be found using the Granger causality. This complies the outcome explained in section 3.6 as only causal

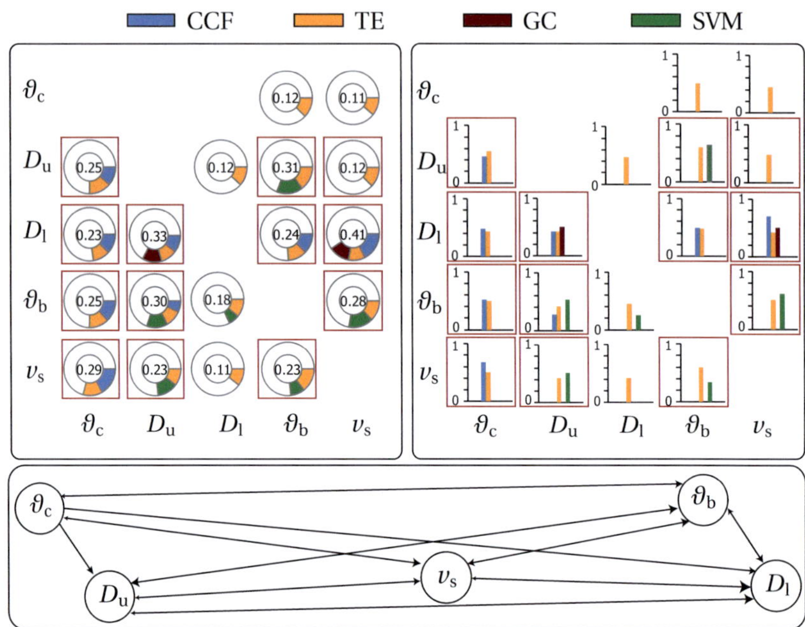

Figure (6.3) Causal matrices of the first production. The left plot shows the combined causal matrix, the plot on the right hand side displays each method in terms of a bar chart. The resulting disturbance propagation path is illustrated in the lower plot.

dependencies containing short dead times can be found because time delays need to be covered by the selected internal order n of the autoregressive model. As explained in section 3.6.1 the maximum model order n_{max} is set to 10 samples which does not cover the dead time from ϑ_{c} to the other process variables. The same counts for the **SVM** as in that case the maximum model order is set to 10 samples (see section 3.7.2). In that case no causal dependency from ϑ_{c} pointing towards the other process variables could be found as the dead time of a system also needs to be covered through the order of n. Additionally, the SVM detects the correct causal dependencies $D_{\mathrm{u}} \to v_{\mathrm{s}}$, $D_{\mathrm{u}} \to \vartheta_{\mathrm{b}}$, $\vartheta_{\mathrm{b}} \to v_{\mathrm{s}}$, $\vartheta_{\mathrm{b}} \to D_{\mathrm{u}}$ and $v_{\mathrm{s}} \to \vartheta_{\mathrm{b}}$. Furthermore, the false causal dependency $D_{\mathrm{l}} \to \vartheta_{\mathrm{b}}$ is found. Finally the result of the first production shows that the CCF works best on the data set. The main reason is that this method is especially suited to deal with large dead times in a data set.

Table 6.2 outlines the root cause priority list which is calculated from the combined causal matrix. The cylinder temperature ϑ_c is correctly detected as variable which detects the disturbance first as it is ranked on first position. The lower diameter D_l is ranked last which marks it as the variable detecting the disturbance last.

Rank	Process variable	RC
1	Cylinder temperature ϑ_c	1.02
2	Pulling speed v_s	0.92
3	Beam temperature ϑ_b	0.9
4	Upper diameter D_u	0.86
5	Lower diameter D_l	0.41

Table (6.2) Root cause priority list of the first production. The cylinder temperature ϑ_c is correctly detected as root cause, the lower diameter is declared as being the output variable detecting the disturbance last.

The same results can be confounded from figure 6.3 when looking at the directed graph. Only two weak causal dependencies point towards ϑ_c which mark this variable as input. In addition, ϑ_c shows a strong causal impact on all the other variables which is due to the fact that transfer entropy and cross-correlation found causal dependencies pointing from this variable. From the graph, the process variable D_l can be interpreted as output variable since $\vartheta_c, \vartheta_b, D_u$ and v_s have a causal impact on it and only weak causal dependencies point from D_l to the other variables. In addition, the graph illustrates the expected intercausal dependencies of the three process variables ϑ_b, v_s and D_u.

6.2.2 Analysis of the Second Production

The measurement data set of the second production, given in figure 6.4, is used for the calculation of the disturbance propagation path. The results are illustrated in figure 6.5. In the doughnut and the bar chart the sketched red squares are used to represent the known causal dependencies from the process knowledge (see figure 6.1). The disturbance propagation path is reconstructed by using all methods and merged to the final causal matrix Q_{fus}. For all methods,

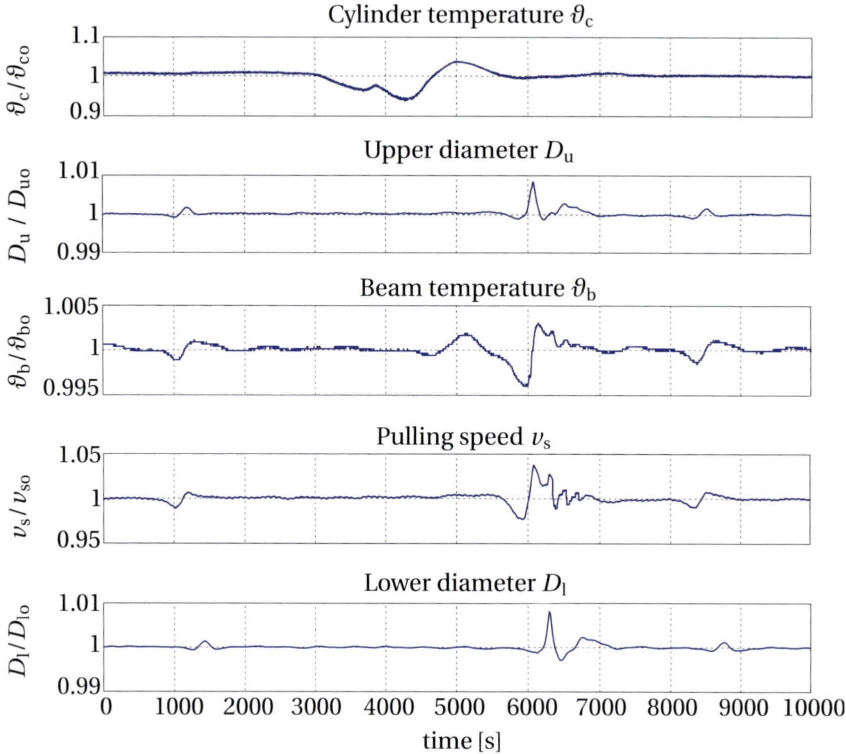

Figure (6.4) Data of the industrial glass forming process containing a weld disturbance. The data is used as the second test for the reconstruction of the disturbance propagation path. In this production the weld disturbance is less distinct in time compared to the first production.

the mean causal matrix $\overline{Q}$ is set to 0.5 and the fitting parameters result in $\beta_\mathrm{CCF} = 0.77$, $\beta_\mathrm{TE} = 0.24$, $\beta_\mathrm{GC} = 0.19$ and $\beta_\mathrm{SVM} = 0.6$.

Comparing the methods by using the generated bar chart, the results show, like for the first production, that the **CCF** leads to the best outcome. No false causal dependencies are found when using this method and only the causal dependencies $\vartheta_\mathrm{b} \rightarrow D_\mathrm{u}$, $v_\mathrm{s} \rightarrow \vartheta_\mathrm{b}$, $\vartheta_\mathrm{b} \rightarrow v_\mathrm{s}$ and $v_\mathrm{s} \rightarrow D_\mathrm{u}$ are not detected. The explanation is like for the first production that the CCF works well on systems with large dead times which is the case for this system.

Compared to the first run, the **TE** shows much better results as only the causal dependency $\vartheta_b \rightarrow \vartheta_c$ has been detected wrong. Furthermore, seven out of thirteen expected causal dependencies are found correctly when applying this method.

Using **GC** yields also more causal dependencies compared to the first run. In addition, it is the only method that detects the causal dependencies $\vartheta_b \rightarrow D_u$ and $v_s \rightarrow D_u$. Still no cause-effect relationship regarding ϑ_c can be found, which can again be explained due to the limited ability when using Granger causality on systems with a large dead time.

Like Granger causality the suggested method based on the **SVM** detects more causal dependencies compared to the first run. As a drawback it needs to be said that both, SVM and GC detect the false cause-effect relationship $D_l \rightarrow v_s$, which results in a strong wrong causal dependency in Q_{fus}.

Compared to the first production, the resulting combined causal matrix of the second run gives a similar picture of the disturbance propagation path. This shows that the results are better than initially expected. As a drawback, the false causal dependency $D_l \rightarrow v_s$ can lead to a wrong interpretation of the path. Since the distortion through the weld is stronger, the time differences are less distinct, which should have resulted in less found causal dependencies. Still, even that v_s is still fluctuating while the weld distortion is already measured at the lower diameter, D_l is correctly detected as output variable. Additionally, it can be concluded that the cylinder temperature ϑ_c is correctly detected as an input variable as there are no strong causal dependencies pointing towards it. Table 6.5 outlines the root cause priority list which is calculated from the combined causal matrix. Like for the first production the cylinder temperature ϑ_c is correctly detected as being the root cause variable meaning that it detects the disturbance first. Still, for this production the resulting value of the root caus priority list for ϑ_c is almost equal to the value for the upper diameter D_u. Hence, the cylinder temperature is not detected as root cause variable without any doubt. Regarding the lower diameter D_l, the results are much clearer. In that case it can doubtlessly concluded, that D_l acts as output variable of the investigated process devices.

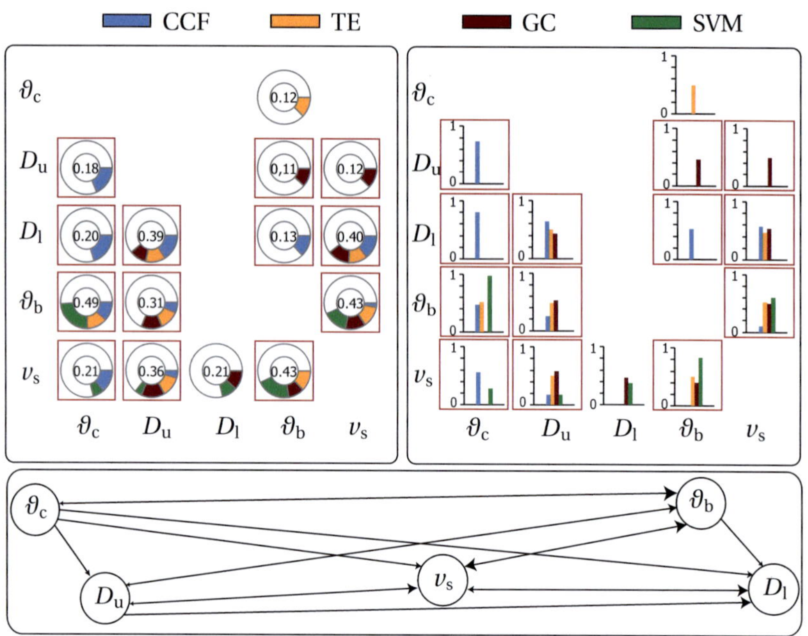

Figure (6.5) Calculated causal matrices of the second production. The left plot shows the combined causal matrix, the plot on the right hand side displays each method in terms of a bar chart. The resulting disturbance propagation path is given in the lower plot.

Rank	Process variable	RC
1	Cylinder temperature ϑ_c	1.08
2	Upper diameter D_u	1.06
3	Pulling speed v_s	0.95
4	Beam temperature ϑ_b	0.79
5	Lower diameter D_l	0.21

Table (6.3) Root cause priority list of the second production. Like in the first production, the cylinder temperature ϑ_c is correctly detected as root cause, the lower diameter is detected as being the output variable.

6.3 Summary

The aim of this chapter was to investigate the behavior of the proposed methods on a data set coming from an industrial plant. To this aim, exemplarily two measurement data sets of known disturbances from a glass forming process were used. The outcome shows that it is possible to use the methods for the reconstruction of the disturbance propagation path as in both productions the causal matrices showed the expected cause-effect relationships known from process knowledge in principle. In detail, ϑ_c was in both productions correctly identified as being the input variable and D_l as output variable.

As the process contains a large dead time for the cylinder temperature ϑ_c and the other process variables, especially the CCF works well on the data sets. The TE faces problems in the first run as it detects many wrong causal dependencies due to the narrow bandwidth in the data, but yields good result in the second run. GC and SVM detect less causal dependencies as both methods face problems if the process contains large dead times, which is the case for this system. In this case both methods cannot cover the dead times in the system due to the choice of a too low model order. Therefore, in the first run neither GC nor the SVM can detect the underlying cause-effect dependencies containing the cylinder temperature ϑ_c.

Another difficulty of the investigated glass forming process is that the process is not running in feed-forward control. Therefore, the ground truth of the disturbance propagation path is in fact a partially directed graph as the system contains a feedback loop. This leads to intercausal dependencies in the measurement data. The feedback loop was correctly detected when combining the methods to one causal matrix.

To summarize, the results show that for the special case of the industrial glass forming process, the suggested methods can reconstruct the fault propagation path of a weld disturbance which again shows that the methods can be applied to on industrial process data.

7

Setup of Data Acquisition System

Usually the ability to reconstruct the disturbance propagation path does not only depend on the occurred fault and the characteristics of the underlying plant but also on the setup of the control and measurement system. Hence, the purpose of this chapter is to check how the suggested methods behave depending on the configuration of the underlying data acquisition. In detail, the detection of cause-effect dependencies depending on the selected sampling period and the data compression rate for the raw signal is investigated. In the last section, the effect of filtering in terms of removing a sinusoidal oscillation in the measurement data is analyzed. As sinusoids are cyclic they do not contain any causal information and only the superposed noise can be used for analysis. Therefore, it is investigated if a suppression of sinusoids increases the ability to detect causal dependencies.

7.1 Sampling Period

The acquired signals used for the reconstruction of the disturbance propagation path are sampled beforehand through the data acquisition system. Here, it is very important to select a proper sampling rate to be able to use the data for causal analysis. If the sampling frequency is set too low it is not possible to reconstruct the continuous-time signal and to cover the dynamics of the system (e. g. if the Nyquist theorem is violated [Joh93]). This can lead to false causal dependencies in the reconstruction of the disturbance propagation path and possibly the wrong root cause is detected. Setting the sampling rate too high leads to an increased load on the data acquisition network and the measurement data possibly needs to be compressed which can have again an impact on the detection of causal dependencies. This will be outlined in section 7.2. Nevertheless, even if the data is not compressed, a chosen sampling frequency that is inappropriate can have a negative impact on the detection of causal dependencies.

In order to determine the impact of different sample periods, the methods are tested on the 1st-order system and the 2nd-order system described in section 3.2 as part of the benchmark data set. The measurements from the two scenarios are acquired using the sampling periods $T_s \in \{0.005\,s, 0.01\,s, 0.025\,s, 0.05\,s, 0.1\,s, 0.15\,s, 0.2\,s, 0.25\,s\}$ and white noise with $\mathcal{N}(0,1)$ is used as input signal. Since the time constant for both system is set to $T = 0.5\,s$, for both system $T_{sn} = 0.25\,s$ defines the limit of the Nyquist sampling theorem.

Initially, this leads to the question which sample rate should be selected for the investigated scenarios.

Åström [AW97] suggests to set the sampling rate in terms of the number of samples per rise time. Therefore, he defines the samples per rise time N_r as

$$N_r := \frac{T_r}{T_s}, \tag{7.1}$$

with T_r being the rise time of the system. As a rule of thumb he proposes to set N_r between 4 and 10. For the 1st-order system T_r is equal to the time constant of $T = 0.5\,s$ which implies that sampling periods of $0.05\,s \leq T_s \leq 0.125\,s$ should give the best results for the detection of causal dependencies.

According to [AW97], for a 2nd-order system, the rise time is calculated as

$$T_r = \frac{T}{2\pi} e^{\phi/\tan\phi}, \tag{7.2}$$

with the damping $\zeta = \cos\phi$. For the investigated 2nd-order system ($\zeta = 0.7, T = 0.5\,s$) this results in the sampling periods of $0.017\,s \leq T_s \leq 0.043\,s$. For both scenarios, it is expected that the methods will give reasonable results in the defined ranges. To test the methods, in each case $K = 10000$ samples are generated and since each method is tested for itself, the fitting parameters $\beta_{CCF}, \beta_{TE}, \beta_{GC}$ and β_{SVM} are all set to 1.

T_s in s	Q^{CCF}		Q^{TE}		Q^{GC}		Q^{SVM}	
	$u \to y$	$y \to u$	$u \to y$	$y \to u$	$u \to y$	$y \to u$	$u \to y$	$y \to u$
0.005	0.86	–	0.036	–	0.49	–	0.27	–
0.01	0.86	–	0.036	–	0.49	–	0.28	–
0.025	0.86	–	0.038	–	0.49	–	0.30	–
0.05	0.86	–	0.038	–	0.49	–	0.28	–
0.1	0.86	–	0.040	–	0.49	–	0.32	–
0.15	0.69	–	0.025	–	0.13	–	0.14	–
0.2	0.79	–	0.030	0.041	0.18	0.01	0.12	–
0.25	0.52	–	0.035	0.031	0.06	0.01	0.16	–

Table (7.1) Causal strengths for the 1st-order system with white noise as input signal and different sampling periods.

T_s in s	Q^{CCF}		Q^{TE}		Q^{GC}		Q^{SVM}	
	$u \to y$	$y \to u$	$u \to y$	$y \to u$	$u \to y$	$y \to u$	$u \to y$	$y \to u$
0.005	0.83	–	0.019	–	0.48	–	0.26	–
0.01	0.84	–	0.020	–	0.49	–	0.33	–
0.025	0.84	–	0.023	–	0.49	–	0.33	–
0.05	0.84	–	0.025	–	0.49	–	0.23	–
0.1	0.85	–	0.041	–	0.50	–	0.26	–
0.15	0.78	–	0.034	–	0.18	–	0.27	–
0.2	0.73	–	0.042	0.052	0.21	0.02	0.21	–
0.25	0.75	–	0.035	0.051	0.13	0.01	0.22	–

Table (7.2) Calculated causal strengths for the 2nd-order system with white noise as input signal and different sampling periods.

Impact of sampling period on test scenarios   The results are outlined in table 7.1 for the 1st-order system and in table 7.2 for the 2nd-order system. In both scenarios the sampling period has an impact on the resulting causal dependencies for all methods. Therefore, the results with short sampling periods show a much lower impact on the causal strengths of the dependency $u \to y$ as the causal strengths only change moderately.

If the sampling period is increased, the impact is stronger, as the causal dependencies result in weak causal strengths and for some methods wrong causal dependencies pointing from $y \rightarrow u$ are detected. All methods yield reasonable values for the proposed intervals from Åström, which is fulfilled for the 1st-order system for the investigated sampling periods $T_s \in \{0.05\,s, 0.1\,s\}$ and for the 2nd-order system for $T_s = 0.025\,s$.

Comparing the methods regarding the impact of the different selected sample periods yields as one of the most robust methods the **CCF**. For both scenarios and all sampling periods the causal dependencies are correctly detected. Still, if the sampling period approaches the limit of the Nyquist sampling theorem, which is for both systems $T_{sn} = 0.25\,s$, the strengths of the causal dependencies are weakened.

The **TE** faces difficulties if the sampling period is set too large as for both scenarios for $T_s \in \{0.2\,s, 0.25\,s\}$ a false causal dependency $y \rightarrow u$ is found. Selecting a short sampling period has no impact for the 1st-order system , for the 2nd-order system the causal strength of the dependency $u \rightarrow y$ weakens.

In both scenarios the **GC** correctly detects for all cases the cause-effect relationship $u \rightarrow y$. Still, if the sampling period approaches the limit of the Nyquist sampling theorem ($T_{sn} = 0.25\,s$), the calculated causal strengths for $u \rightarrow y$ weaken and additionally a weak causal dependency pointing from $y \rightarrow u$ is detected.

Like the CCF, the **SVM** is robust against the selected sampling period as in none of the cases a cause-effect relationship $y \rightarrow u$ has been found. If the sampling period is long, significant causal dependencies can be found with a weak causal strength. The short sampling periods have no impact on the causal strength as the values stay almost constant for all selected values of T_s.

In summary, if the sampling period for the analyzed system is set to a short value, some methods yield lower causals strength but the correct causal direction is found. Selecting a too large sampling period can result in significant wrong causal dependencies. In that case, the reconstructed disturbance propagation path of the fault can be incorrect due to the false causal dependencies. Finally, it can be concluded that if the sampling period is selected reasonably the methods give meaningful results. Selecting a reasonable sampling period can be achieved by following the recommendations from Åström which are given in equation 7.1 for 1st-order and in equation 7.2 for 2nd-order systems. Both recommendations take into account the time constant of the underlying system which is analyzed.

7.2 Impact of Lossy Data Compression

In some cases, long time series need to be compressed from the data acquisition system to reduce the data storage space. Generally, lossy data compression means that not the complete data set but only fractions of it are kept in the database. Since some of the data is discarded, this has an impact on the detection of causal dependencies. In recent literature there are several possible ways described on how to compress time series and a good overview is given e. g. in [LKB04]. Since the details of the different types of data compression are out of scope they are not discussed. In order to determine the impact of lossy data compression for the detection of causal dependencies an approach called piecewise constant approximation (PICA) is used. The concept of the PICA is that each measured time series is divided into equal segments and the mean value of each segment is then used to reconstruct the compressed time series used for analysis. In recent literature there exist several investigations on the compression rate using this method while e. g. Chakrabarti [KESM02] shows that rates between 10 to 60 times of the original length of the time series can be reached with this method while still having a reasonable loss of information. Figure 7.1 gives an illustration of the described method.

In that case the compression rate was set to $c = 5$, which means that the original time series is compressed by a factor five. The original signal is displayed in the upper plot and the resulting compressed time series is given in the lower plot. In order to test the effect of data compression on the methods, the 1st-order system, which was used as the base configuration described in section 3.2, is again utilized. In the first case white noise with $\mathcal{N}(0,1)$ is selected as input signal. For the data acquisition a sample period of $T_s = 0.1\,\text{s}$ is selected. In the second case the white noise is filtered beforehand using a low-pass filter with a time constant of $T = 5\,\text{s}$ to generate an input signal with a limited bandwidth. In total, four different compressions rates, namely $c \in \{2, 5, 10, 15\}$ are used while having $K = 10000$ samples in the uncompressed case. As each method is tested separately, the fitting parameters β_{CCF}, β_{TE}, β_{GC} and β_{SVM} are all set to 1.

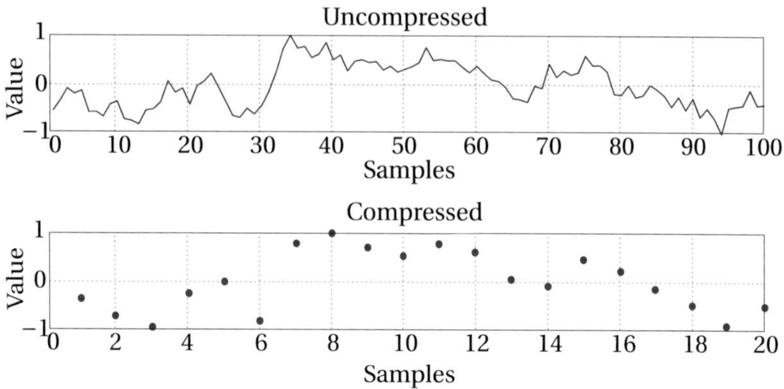

Figure (7.1) Illustration of a compressed time series. The upper plot shows acquired measurement data from the control system, the lower plot shows the compressed signal while using a compression rate of $c = 5$.

Results using white noise as input signal Table 7.3 shows the calculated causal strengths for diverse compression rates in the presence of white noise as input signal. As the 1st-order system is the base configuration from the benchmark data, in the uncompressed case the causal strengths are the same as in chapter 3. The results indicate that all methods are affected by the compression of the time series since on average the causal strengths get lower with an increasing compression or the causal dependency is no longer detected as being significant.

The results for all investigated methods show that the **CCF** is strongly affected by compression as it can only detect cause-effect relationships up to a rate of $c = 2$ which makes this method the least robust one. In the other cases, the causal dependency detected by the CCF is considered as being not significant.

Compared to the CCF, the **TE** yields more robust values when having compressed data at hand as up to a compression rate of $c = 10$ the causal dependencies are found correctly. Furthermore, the calculated causal strength does not change relevantly with regard to the different compression rates.

The outcome of the **GC** shows that it is the most robust method since it is the only method which detects the cause-effect relationships up to a rate of $c = 15$. In all cases, the causal dependencies are found while the values show a reduction of the causal strength with an increasing compression rate.

The **SVM** detects causal dependencies up to a rate of $c = 5$, but the causal strength is close to zero. In other words, like the CCF, the SVM is strongly influenced by data compression when it is used for the calculation of the fault propagation path.

	Q^{CCF}		Q^{TE}		Q^{GC}		Q^{SVM}	
	$u \to y$	$y \to u$	$u \to y$	$y \to u$	$u \to y$	$y \to u$	$u \to y$	$y \to u$
Uncompressed	0.86	–	0.04	–	0.49	–	0.32	–
$c = 2$	0.52	–	0.06	–	0.45	–	0.49	–
$c = 5$	–	–	0.03	–	0.42	–	0.05	–
$c = 10$	–	–	0.03	–	0.27	–	–	–
$c = 15$	–	–	–	–	0.12	–	–	–

Table (7.3) Causal strengths for the base configuration of the benchmarks depending on the selected compression rate while having white noise as input signal.

Results using colored noise as input signal　The results when using noise with a limited bandwidth as input data is given in table 7.4. In this case, the white noise is filtered beforehand using a low-pass filter with a time constant of $T = 5\,$s. As explained in chapter 3 all methods face problems with the reduction of the bandwidth for the input signal. Therefore, it is expected that the proposed methods will have more difficulties to detect the causal dependency. Compared to table 7.3 in which white noise was used for compression, the results show that the found causal dependencies in this case are less clear. In addition, false causal dependencies pointing from $y \to u$ have been found for some methods.

Analyzing the results of the methods in further detail, starting with the **CCF**, this reveals that no causal dependency could be found even when the data set is left uncompressed. This is due to the fact that the CCF is the one of the less robust analyzed methods regarding autocorrelation in the input signal for this 1st-order system.

The **TE** shows similar results for the data with limited bandwidth as if white noise was used as input signal. In this case, up to a compression rate of $c = 10$ the cause-effect relationship is detected as being significant. Additionally, no trend in the causal strengths can be found with regard to the selected c.

The **GC** detects for all compression rates the causal dependency $u \to y$ with a weaker value for the causal strength with an increasing compression rate. In addition, for the high compression rates $c \in \{10, 15\}$ a false causal dependency pointing from $y \to u$ with a low causal strength is found.

A similar result counts for the **SVM**. Up to a compression rate of $c = 10$ the cause-effect relationship $u \to y$ is found, but for $c \in \{5, 10\}$ the method also detects the false causal dependency $y \to u$ with a weak causal strength.

	Q^{CCF}		Q^{TE}		Q^{GC}		Q^{SVM}	
	$u \to y$	$y \to u$	$u \to y$	$y \to u$	$u \to y$	$y \to u$	$u \to y$	$y \to u$
Uncompressed	–	–	0.04	–	0.40	–	0.28	–
$c = 2$	–	–	0.04	–	0.50	–	0.33	–
$c = 5$	–	–	0.06	–	0.48	–	0.02	0.03
$c = 10$	–	–	0.06	–	0.27	0.02	0.10	0.01
$c = 15$	–	–	–	–	0.15	0.04	–	–

Table (7.4) Causal strengths for the 1st-order system depending on the selected compression rate when having colored noise as input signal.

Summary of the results   In this section, exemplarily the impact of a lossy data compression for data acquisition was checked. In detail, the impact when having white and colored noise on a 1st-order system was investigated. The results illustrate that already compression rates with a small loss of data have an impact on the detection of causal dependencies in a data set. In the case of white noise the causal dependencies for high compression rates, which means a high loss of information, are not found as being significant or they result in low causal strengths. Regarding the data with colored noise, the outcome gives similar results. Furthermore, in the case of colored noise, a high compression rate can lead to the detection of false causal dependencies. Hence, it can be concluded that if the proposed methods for the calculation of the disturbance propagation path are used, a lossy compression of the data should be avoided since this can have a strong impact on the reconstruction of the disturbance propagation path and finally the detection of the root cause of a fault.

7.3 Filtering Oscillations

This section deals with the problem of oscillations and the impact of filtering these oscillations as a preprocessing step when calculating causal dependencies. In this context, exemplarily sinusoidal oscillations are selected. As sinusoidal oscillations are deterministic they do not incorporate causal information. This becomes obvious if two signals $u(t)$ and $y(t)$ are observed which are of oscillatory nature with the same periodicity but at a different phase. In that case both signals could be seen as being the cause of the other. Therefore, the only way to detect cause-effect relationships among these variables is included in the information contained in the noise superposed to the oscillation. Hence, it is expected that if in a preprocessing step the main oscillations in the data are removed, hidden causal information can be better revealed. In this section the impact on the proposed methods when having the special case of removing the main oscillation by using a notch filter is analyzed. In a similar case this has been tested for Granger causality by Florin [FGP$^+$10] who added a sinusoid with a negative sign on a signal to remove the 50 Hz current in a Magnetoencephalographic signal. He showed that using this approach gives better results than keeping the oscillation in the acquired measurements. A similar result for Granger causality is given by Barnett [BS11] who also removed electrical line noise from the acquired data set.

Notch filter as a band-stop   Notch filters are filters that are used to remove a particular frequency component while not affecting significantly nearby frequencies. Hence, they can be considered as a special type of a band-stop filter. There are several ways of how to design notch filters and for an overview it is referred to [Sch10]. In the present work a standard notch filter is used which

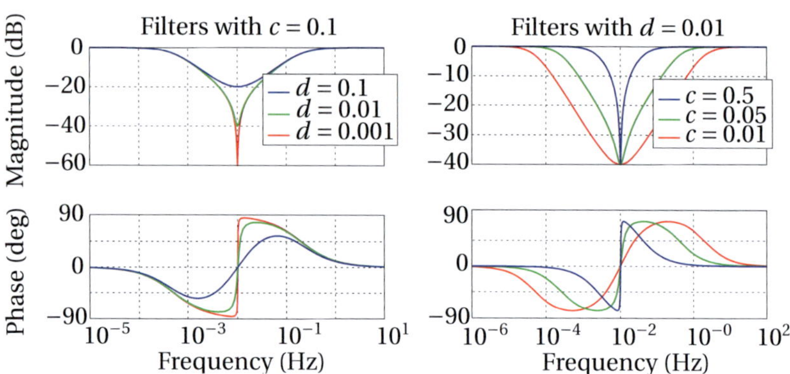

Figure (7.2) Resulting Bode plots showing the phase and the magnitude of the transfer function of the notch filter with differently selected values for the parameters c and d. The notch frequency ω_n has been set to 10^{-2} Hz.

has three parameters, namely the selection of the notch frequency ω_n and two tuning parameters d and c. This results for the transfer function $H(s)$ in

$$H(s) = \frac{s^2 + 2\frac{d}{c}\omega_n + \omega_n^2}{s^2 + \frac{2}{c}\omega_n + \omega_n^2}. \tag{7.3}$$

The parameter d is used to define the damping of the notch frequency and the parameter c can be considered as the sharpening parameter in terms of the nearby frequencies.

Figure 7.2 shows Bode plots for different selections of the filter parameters d and c which will be used for analysis.

Filtering oscillatory data using a notch filter can be seen as a trade-off. If the oscillations are not completely suppressed there can still be some information hidden in the signal which cannot be detected by the proposed methods. On the other hand, if the filtering is set to sharp, Barnett [BS11] shows that artifacts can occur in the data that possibly lead to false cause-effect relationships. In order to test the influence of the detection of causal dependencies when having oscillations in data and when filtering them, the base configuration from the benchmarks (section 3.2) is used. As input a sinusoidal signal with the frequency $f = 10^{-2}$ Hz and an amplitude $\hat{u} = 1$ is used while being superposed with white noise consisting of $\mathcal{N}(0, 1)$. The data is sampled with $T_s = 0.1$ s. The investigated values of the different filter parameters c and d correspond to the ones shown in figure 7.2 and are used to remove the oscillations from the input and the output signal. Since each method is tested for itself all fitting parameters β_{CCF}, β_{TE}, β_{GC} and β_{SVM} are set to a value of 1.

	Q^{CCF}		Q^{TE}		Q^{GC}		Q^{SVM}	
	$u \to y$	$y \to u$	$u \to y$	$y \to u$	$u \to y$	$y \to u$	$u \to y$	$y \to u$
No oscillations	0.86	–	0.04	–	0.49	–	0.32	–
Non-filtered	–	–	–	–	0.95	0.18	–	0.03
$c = 0.1, d = 0.1$	–	–	0.03	0.12	0.95	0.21	0.26	0.14
$c = 0.1, d = 0.01$	0.57	–	0.24	–	1.00	0.03	0.91	–
$c = 0.1, d = 0.001$	0.51	–	0.25	–	1.00	0.01	0.17	–
$c = 0.5, d = 0.01$	0.83	–	0.23	–	0.51	–	0.16	–
$c = 0.05, d = 0.01$	0.39	–	0.26	–	1.00	0.06	–	–
$c = 0.01, d = 0.01$	0.26	–	0.14	–	1.00	0.19	–	–

Table (7.5) Causal strengths from the suggested methods when having a sinusoidal oscillating input data on the base configuration from the benchmark data. The parameters c and d are used to tune the notch-filter.

Results of oscillation filtering Table 7.5 shows the outcome of the different methods if a notch filter is applied with varying parameter values for c and d. In addition, the calculated causal strengths for the methods without performing any filtering and if there would be no oscillations in the data are given for comparison. The results reveal that all methods are affected when having oscillations in the data. If the sinusoid is not suppressed, no method detects the correct causal dependency $u \to y$ reliably. As expected, suppressing the sinusoid increases for all methods the probability of detecting the correct causal dependency. But the outcome also shows that filtering needs to be done carefully since only for the parameter setting $c = 0.5$ and $d = 0.01$ all methods work correctly. Intuitively, the results indicate increasing the value of c further. Still, as illustrated in the Bode plot in figure 7.2, setting $c = 0.5$ generates already a sharp notch in the frequency spectrum, which means that the frequency of

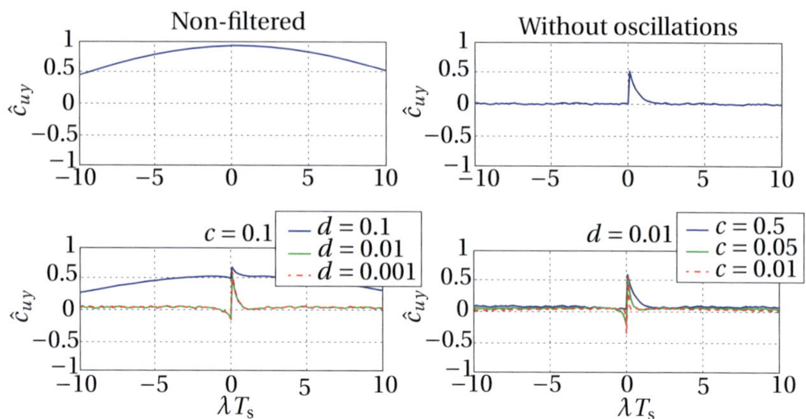

Figure (7.3) Illustration of behavior of the CCF when filtering oscillations in measurement data on a 1st-order system. Several different parameter values c and d are used for tuning the notch-filter. The upper plots represent the non-filtered case and the case without oscillations in the signal.

the sinusoid needs to be known exactly. As a possibility, the detection of the frequency of the sinusoid can be done by performing a spectral analysis of the signal. If the frequency of sinus is shifting it is also possible to adapt the selected notch frequency.

Analyzing the results of each method, the **CCF** shows a much better performance if the notch filter is used. Except for the parameter combination $c = 0.1, d = 0.1$ the causal dependency $u \to y$ is always detected correctly. If the data is not filtered, the causal dependency $u \to y$ is found as being not significant. In order to go further into detail, the impact of filtering is illustrated for the cross-correlation function in figure 7.3. If the data is not filtered, the CCF results in a curve with a flat maximum and no causal information can be drawn from it. If a soft filtering is performed, by setting $c = 0.1$ and $d = 0.1$, only a small peak is present in the CCF which is not detected as being significant by the proposed algorithm. Increasing the sharpening effect by setting the filter parameters to $c = 0.01, d = 0.01$ or $c = 0.05, d = 0.01$ shows the other extreme as an artifact in terms of an inversed peak results when calculating the cross-correlation function. In these two cases the cause-effect relationship is still detected correctly as being significant but gives a lower causal strength due to the occurred inverse peak.

The **TE** implies almost the same performance as the CCF, except for the filter parameters $c = 0.1, d = 0.1$. In that case the transfer entropy detects an intercausal dependency $u \leftrightarrow y$.

The **GC** detects in all cases a strong causal dependency $u \rightarrow y$, but except for $c = 0.5, d = 0.01$, also a weak causal dependency pointing from $y \rightarrow u$ is found. Nevertheless, in all cases the causal strength $u \rightarrow y$ is larger than $y \rightarrow u$ which shows the advantage when calculating the causal strength and not solely relying on significance tests. Additionally, from all proposed methods the GC is the only method detecting $u \rightarrow y$ in the non-filtered case.

The **SVM** does not detect $u \rightarrow y$ if the filter parameters are set to $c = 0.05, d = 0.01$ or $c = 0.01, d = 0.01$. Furthermore, for the parameter setting $c = 0.1, d = 0.1$ a weak false causal dependency $y \rightarrow u$ is found. This underlines that the SVM is the most sensitive method with respect to filtering oscillations in measurement data.

The results show, if the parameters are set appropriately, filtering is a suitable preprocessing step for the detection of causal dependencies. But an incorrect setting of the filter parameters can lead to delusive results since some of the methods detect false causal dependencies. This is the reason, why filtering should only be done, if the oscillation is well known beforehand or if an adative notch filter can be applied.

7.4 Summary

This chapter illustrated that both, the selection of the sampling period and when applying a data compression prior to performing a causal analysis, have an impact on the ability to detect causal dependencies in measurement data. As test process the base configuration and the 2nd-order system of the benchmark data were used.

In terms of the sample time, the results show, that when selecting a too long sampling period, no causal dependencies are found or the detection of false causal dependencies can be the result when applying the proposed methods.

Still, when setting the sampling period of a system to a reasonable value, e. g. by following the recommendation proposed by Åström [AW97], the impact of the sampling period when calculating cause-effect relationships can be neglected. Performing a lossy data compressing means that parts of the information contained in the data is lost when storing it in a data base. As shown, all methods are affected by lossy data compression and even small compression rates can have an impact on the result of the causal analysis. In that case, the main advice when performing a causal analysis is that the compression rate should be reduced to a minimum when performing a lossy data compression.

Finally, this chapter illustrated that filtering oscillations, e. g. by removing 50 Hz supply noise, can help to increase the detection rate of causal dependencies in data. Still, filtering should be handled with care, as this can lead to artifacts in the preprocessed measurement data. This again can lead to the detection of false causal dependencies when using the proposed methods. In that case the advice is that filtering should be done if a spectral analysis can be performed and an adaptive notch filtering is used. If oscillations exist in the data set, but the frequencies cannot be clearly detected using the spectral analysis, filtering should be avoided as this can lead to delusive results when performing a causal analysis.

8

Conclusion

A fault, e. g. generated through an error in a major process device, leads to disturbances in a process. These disturbances can have an effect on the performance of the process, put the safety in the plant on risk or can reduce the overall product quality. Since modern control systems comprise of a large number of process variables, which are interconnected to each other, eventually, a fault generates plant-wide disturbances. Since not all relations of the different process variables are well known, the detection of the source of a fault and the reconstruction of the corresponding disturbance propagation path is not an easy task and often a correct fault diagnosis mainly depends on the experience and knowledge of the process operator. To this end, it is of great benefit to automatically localize the source of a disturbance, commonly called the root cause, and to track back the disturbance propagation path.

A literature review on process supervision was given in chapter 2 which showed that only a small amount of plants is modeled in an analytical way. This means that a fault localization based on physical models of the process is usually not possible and data-driven methods can bring a great benefit. The main idea, when using these methods is to exploit statistical dependencies and time-shifts in the measurement data to reconstruct the disturbance propagation path. Methodologically, this means that cause-effect dependencies among the process variables need to be detected. Process variables that show a strong causal impact on other variables come into consideration as being the root cause.

In chapter 3 four different methods were proposed which can be used to detect cause-effect dependencies in process data. Among them, a new method based on support vector machines with variable selection and model reduction and a considerable improvement of a method which uses the cross-correlation function with permutation tests were presented. Additionally, transfer entropy and Granger causality have been extended by calculating a quantitative value, called causal strength, to make all four methods comparable to each other. In detail, all methods result in a value between 0 (= no cause/effect) and 1 (= sure cause/effect) for each calculated causal dependency.

To evaluate the characteristics of each method, in section 3.2 several benchmark data sets were developed. These benchmarks cover aspects such as large dead times, nonlinearities, feed-through or a set-point change. The results indicate that each method has its advantages and drawbacks depending on the characteristics of the underlying data set.

Hence, to cover the drawbacks of one method, a novel approach to combine all methods to one causal matrix has been proposed in chapter 4. This approach works in two stages. First, performing a parameter fitting, the causal matrices are balanced to each other; second, assuming that each method works equally well on the data set, the mean over all causal matrices is taken to calculate the combined causal matrix. Additionally, in section 4.2.2 an approach is presented which adapts the weight of each method when calculating the combined causal matrix, if some a priori knowledge in terms of known causal dependencies is available.

To proof the concept, in chapter 5 the methods were tested on a laboratory plant that consists of an installation in which water is recirculated in two tanks. The conducted experiments consist of failures, such as a loose electricity connection of the pump, an air pressure leak of the valve and the clogging of an inlet pipe. The outcome revealed that more robust results for the root cause detection can be achieved when following the proposed combination of the methods.

Chapter 6 introduced a new industrial case study. In detail, the functionality of the methods has been verified on an industrial glass forming process when reconstructing the disturbance propagation path of a defect contained in the glass cylinder. The results indicate that it is possible to reconstruct the disturbance propagation path when calculating the combined causal matrix in principle.

The impact of the setup of the data acquisition system for the detection of causal dependencies was studied in chapter 7. The results of this study are established in terms of guidelines for the correct selection of the sampling period, the data compression rate and the filtering of oscillatory data.

To represent the causal matrices in a suitable manner, a new visualization method for the causal matrices was needed and has been developed in section 3.3. The visualization contains the combined causal matrix, a comparison of the causal matrices from every method and the detected disturbance propagation path in terms of a directed graph.

8.1 Investigated Methods for Fault Localization

As previously mentioned, in this work four different methods were presented for the detection of the root cause and the reconstruction of the disturbance propagation path. All methods are based on the calculation of statistical dependencies considering temporal information for the detection of cause-effect relationships. In detail, the used methods were the cross-correlation function, Granger causality, transfer entropy and support vector machines.

Cross-correlation function　　The CCF is a linear measure which calculates the correlation of two signals that are time delayed against each other. In order to transform the CCF into a causal measure the calculation of a compound parameter is proposed. This parameter tests if there is a significant difference between the absolute maximum value for a negative time-lag and the absolute maximum value for a positive time-lag. In order to find significant causal dependencies a 3σ-test based on a permutation test is suggested. The outcome for the benchmark data sets reveals that the CCF is especially suited for linear systems with large dead times. However, the method faces problems if the disturbances are narrow-banded, non-steady or the investigated system contains nonlinearities.

Transfer entropy   The concept of the TE is based on the calculation of transition probabilities. In order to detect causal dependencies, the TE tests if the probability of a future value of one variable is increased by having the knowledge of past values of a second variable compared to having the knowledge if only the first variable would be present. Similar to the CCF a 3σ-test on permutated data is used to check if there is a significant cause-effect relationship. Using the benchmark data sets the results show that the TE is suited to detect disturbances when having large dead times and nonlinearities in the underlying process. However, as drawback, the TE has problems if the input signal is narrow-banded.

Granger causality   The GC is based on the comparison of residuals coming from two models, namely an autoregressive model that contains only past values of the first signal, called the restricted model, and an augmented autoregressive model containing past values of the first and second signal, called the unrestricted model. The detection of the causal dependencies was performed by comparing the residuals of the two models. If the residuals of the second model are significantly lower than the residuals of the first model, it is concluded that a causal dependency exists. In the benchmark data sets, the results illustrate that this method can be used when having non-steady or narrow-banded input signals. Since the underlying autoregressive model is linear, nonlinear causal dependencies cannot be detected reliably. In addition, the method faces problems if the dead time among the investigated signals is too large since the GC performs dead time estimation internally.

Support vector machine   The SVM is proposed as a machine learning method for the calculation of cause-effect relationships in the data. Therefore, the main concept is to train and fit the parameters of the SVM based on a one-step-ahead prediction of the first signal using past values of the first and the second signal. As a next step, a dimensionality reduction of the SVM input variables is performed by selecting only input data that has the most relevant information for predicting future values of the first signal. If past values of the second signal are kept in the reduced model, this means that a cause-effect relationship from the second signal to the first one is detected. The benchmark data sets indicate that the SVM is especially suited when the underlying process contains nonlinear dependencies or if the disturbance is non-steady. However, the SVM shows problems if the disturbance is narrow-banded or if the dead time among the investigated signals is too large as the dead time is estimated internally.

8.2 Combination of the Methods

As all methods show benefits and drawbacks on the benchmark data sets, the causal matrices of the methods are combined into one resulting causal matrix to achieve a more robust outcome. Hence, the causal matrices from each method are balanced by following the assumption that all methods work equally well on the investigated data set for the found causal dependencies.

This approach was tested on a continuous stirred tank reactor, on a laboratory plant as well as on an industrial glass forming process. The results give evidence that the combination of the methods brings a large benefit as causal dependencies falsely detected by one method can be compensated through the other methods. Furthermore, the causal matrix is transformed into a root cause priority list, containing a ranking of the process variables being most probable the root cause of the disturbance. With the exception of the experiment that contains an oscillating pump, in all conducted experiments on the laboratory plant the variable responsible for the root cause was ranked first. In the experiment with the oscillating pump the actual root cause was ranked second. Regarding the continuous stirred tank reactor, where two causes are acting on the system, the variables causing the disturbances were ranked first and second. In order to gain further insights into the resulting combined causal matrix, several visualization techniques are proposed.

Visualization of the causal matrices When calculating cause-effect relationships, all methods return their own detected disturbance propagation path in form of a causal matrix. For the visualization of the different matrices, several aspects have to be considered. In detail, an easily understandable representation of the fault propagation path, a comparison of the methods used on the same data set and an appropriate representation of the combined causal matrix are required. In this thesis a method has been developed which addresses all of these aspects. Therefore, each causal matrix is represented in terms of a bar chart, a doughnut chart and the combined causal matrix as a partially direct graph. The bar chart is used to give a comparison of the different methods wherein 0 (= no cause/effect) is defined as the minimum value of the bar chart and 1 (= certain cause/effect) is defined as the maximum. The doughnut chart is used to aggregate the outcome of the different methods to the combined causal matrix wherein the arc length of the doughnut represents the resulting causal strengths of each method. Finally, the partially directed graph is used to represent the disturbance propagation path, wherein the nodes represent the process variables and the edges the cause-effect relationships. The developed visualization method leads to a fast comprehension of the causal dependencies contained in the data set of the investigated processes. Still, for larger data sets an interactive visual interface is desirable for analysis.

8.3 Future Work

Although this work has opened a way to perform a data analysis for the detection of causal dependencies in time series data sets, many open issues for future research remain. Several of which that arose during the work on this thesis are discussed below.

Concerning the acquired data, a set of questions addresses the behavior of the methods depending on the data quality. The performance of the methods to calculate the disturbance propagation path has not been investigated with respect to long-term trends, time-variant system behavior or unknown aliasing effects in the data. Furthermore, preprocessing the data set seems to play a crucial role when detecting causal dependencies.

Another open issue is integrating a priori knowledge adequately in the methods since operators usually have a thorough understanding of the plant behavior. Incorporating this a priori knowledge was outlined as an approach in section 4.2.2. In that case a method selection was performed depending on the known causal dependencies. However a priori knowledge can also be applied to tune a method itself and does not necessarily need to be used only when combining the methods. In this context appropriate concepts could be developed to make more use of the available process knowledge.

The application of the presented methods can be expanded. There are other domains imaginable in which the suggested algorithms can be applied. A first domain is **biosignal analysis**. Massive amounts of data are acquired by measuring biological activities in terms of electric signals and the knowledge of the causal flow is not always obvious. The proposed methods can be used for the analysis of brain signals or muscle activity. Another possible application is the propagation analysis of a contamination in **water distribution networks**. By knowing the cause-effect relationships among the different process variables in the network, the distribution of the contamination can be analyzed and possible counteractions can be undertaken (e. g. by closing sliders to stop the contamination). Using the proposed methods is also possible in the area of **video analysis** as the proposed algorithms can be used to identify temporal interaction patterns occurring in a video (e. g. analyzing behavior of offense/defense in team sport). Additionally, there are many other domains imaginable. Further examples are **traffic analysis** (e. g. detecting the cause of a traffic jam), **security monitoring of major events** (e. g. analysis of the cause of a panic), **risk analysis for insurances** or the analysis of **economic data**.

Bibliography

[ABR64] AIZERMAN, M. A. ; BRAVERMAN, E. A. ; ROZONOER, L.: Theoretical foundations of the potential function method in pattern recognition learning. In: *Automation and Remote Control*, 1964, S. 821–837

[Aka74] AKAIKE, H.: A new look at the statistical model identification. In: *IEEE Transactions on Automatic Control* 19 (1974), December, Nr. 6, S. 716–723. – ISSN 0018–9286

[AMM90] ALEKSANDAR, C. ; MIODRAG, J. ; MILAN, M.: Analytical first-order dynamic model of binary distillation column. In: *Chemical Engineering Science* 45 (1990), Nr. 12, S. 3585–3592

[Ari99] ARIS, R.: *Mathematical Modeling: A Chemical Engineer's Perspective*. Academic Press, 1999 (Process Systems Engineering)

[AW97] ASTROEM, K. J. ; WITTENMARK, B.: *Computer-controlled systems (3rd ed.)*. Prentice-Hall, Inc., 1997. – ISBN 0–13–314899–8

[BA02] BURNHAM, K. ; ANDERSON, D.: *Model selection and multimodel inference: a practical information-theoretic approach*. 2nd. Springer, 2002. – ISBN 0387953647

[Bau05] BAUER, M.: *Data-driven Methods for Process Analyis*, University College London, Diss., 2005

[BFSS03] BYVATOV, E. ; FECHNER, U. ; SADOWSKI, J. ; SCHNEIDER, G.: Comparison of Support Vector Machine and Artificial Neural Network

Systems for Drug/Nondrug Classification. In: *J. Chem. Inf. Comput. Sci.* 43 (2003), S. 1882–1889

[BGV92] BOSER, B. E. ; GUYON, I. M. ; VAPNIK, V. N.: A training algorithm for optimal margin classifiers. In: *Proceedings of the fifth annual workshop on Computational learning theory*, 1992 (COLT '92). – ISBN 0–89791–497–X, S. 144–152

[BS11] BARNETT, L. ; SETH, A. K.: Behaviour of Granger causality under filtering: Theoretical invariance and practical application. In: *Journal of Neuroscience Methods* 201 (2011), Nr. 2, S. 404–419. – ISSN 01650270

[CB00] COUVREUR, C. ; BRESLER, Y.: On the Optimality of the Backward Greedy Algorithm for the Subset Selection Problem. In: *SIAM J. Matrix Anal. Appl.* 21 (2000), Februar, Nr. 3, S. 797–808. – ISSN 0895–4798

[CL05] CHANG, M. ; LIN, C.: Leave-one-out Bounds for Support Vector Regression Model Selection. In: *Neural Computation* 17 (2005), S. 1188–1222

[CMLVQ03] CHÁVEZ, M. ; MARTINERIE, J. ; LE VAN QUYEN, M.: Statistical assessment of nonlinear causality: application to epileptic EEG signals. In: *Journal of Neuroscience Methods* 124 (2003), Nr. 2, S. 113–128

[CRB01] CHIANG, L.H. ; RUSSELL, E. ; BRAATZ, R. D.: *Fault Detection and Diagnosis in Industrial Systems*. Springer, 2001 (Advanced Textbooks in Control and Signal Processing). – ISBN 9781852333270

[CST00] CRISTIANINI, N. ; SHAWE-TAYLOR, J.: *An Introduction to Support Vector Machines and Other Kernel-based Learning Methods*. 1. Cambridge University Press, 2000. – ISBN 0521780195

[CT06] COVER, Thomas M. ; THOMAS, Joy A.: *Elements of Information Theory (Wiley Series in Telecommunications and Signal Processing)*. Wiley-Interscience, 2006. – ISBN 0471241954

[DCB06] DING, M. ; CHEN, Y. ; BRESSLER, S. L.: Granger Causality: Basic Theory and Application to Neuroscience. (2006), August

[DHS01b] DUDA, R. O. ; HART, P. E. ; STORK, D. G.: *Pattern classification*. Wiley, 2001

[DM98] DESBOROUGH, L. ; MILLER, R.: *Increasing customer value of industrial control performance monitoring-Honeywells experience*. New York American Institute of Chemical Engineers, 1998. – 169–189 S

[DOT03] DOUKHAN, P. ; OPPENHEIM, G. ; TAQU, M. S.: *Theory and Applications of Long-Range Dependence*. Birkhäuser, 2003. – ISBN 9783764341688

[DP05] DING, C. ; PENG, H.: Minimum redundancy feature selection from microarray gene expression data. In: *Journal of Bioinformatics and Computational Biology* 3 (2005), Nr. 2, S. 185–205. – ISSN 0219–7200

[DW50] DURBIN, J. ; WATSON, G. S.: Testing for serial correlation in least squares regression. In: *Biometrika* 37 (1950), Nr. 3-4, S. 409–428

[Efr79] EFRON, B.: Bootstrap Methods: Another Look at the Jackknife. In: *The Annals of Statistics* 7 (1979), Nr. 1, S. 1–26. – ISSN 00905364

[EM07] EATON, D. ; MURPHY, K.: Belief net structure learning from uncertain interventions. In: *Journal of Machine Learning Research* 1 (2007)

[FGP+10] FLORIN, E. ; GROSS, J. ; PFEIFER, J. ; FINK, G. ; TIMMERMANN, L: The effect of filtering on Granger causality based multivariate causality measures. In: *NeuroImage* 50 (2010), Nr. 2, S. 577 – 588. – ISSN 1053–8119

[Föl08] FÖLLINGER, O.: *Regelungstechnik: Einführung in die Methoden und ihre Anwendung.* 10. Huethig, 2008

[Fre08] FREY, C. W.: Diagnosis and Monitoring of Complex Industrial Processes based on Self-Organizing Maps and Watershed Transformations. In: *Computational Intelligence for Measurement Systems and Applications* (2008), S. 87–92

[Fri00] FRIEDMAN, N.: Being Bayesian about network structure. In: *Machine Learning*, 2000, S. 201–210

[Fri04] FRIEDMAN, N.: Inferring Cellular Networks Using Probabilistic Graphical Models. In: *Science* 303 (2004), Nr. 5659, S. 799–805

[GAE06] GUYON, I. ; ALIFERIS, C. ; ELISSEEFF, A.: Causal feature selection. In: *NIPS Workshop on Causality and Feature Selection* (2006)

[GC10] GELPER, S.E.C. ; CROUX, C.: On the Construction of the European Economic Sentiment Indicator. In: *Oxford Bulletin of Economics and Statistics* 72 (2010), Nr. 1, S. 47–62

[GE03] GUYON, I. ; ELISSEEFF, A.: An introduction to variable and feature selection. In: *J. Mach. Learn. Res.* 3 (2003), S. 1157–1182. – ISSN 1532–4435

[Ger98] GERTLER, J.: *Fault Detection and Diagnosis in Engineering Systems.* Marcel Dekker, 1998. – ISBN 9780824794279

[Gla08] GLASMACHERS, T.: *Gradient based optimization of support vector machines*, Ruhr-Universität Bochum, Diss., 2008

[Gly01] GLYMOUR, C.: *The Mind's Arrows: Bayes Nets and Graphical Causal Models in Psychology.* MIT Press, 2001

[Gra69] GRANGER, C. W. J.: Investigating Causal Relations by Econometric Models and Cross-Spectral Methods. In: *Econometrica* 37 (1969), July, Nr. 3, S. 424–38

[Gra10] GRAY, R. M.: *An Introduction to Statistical Signl Processing.* Stanford University, 2010. – ISBN 9780521131827

[GZ05] GENG, Z. ; ZHU, Q.: Multiscale Nonlinear Principal Component Analysis (NLPCA) and Its Application for Chemical Process Monitoring. In: *Industrial & Engineering Chemistry Research* 44 (2005), Nr. 10, S. 3585–3593

[Hal08] HALL, D. B.: Analyzing Receiver Operating Characteristic Curves With SAS. In: *The American Statistician* 62 (2008), Nr. 4, S. 362–362

[HCL03] HSU, C. ; CHANG, C. ; LIN, C.: A Practical Guide to Support Vector Classification. (2003)

[Hei10] HEIMAN, G.W.: *Basic Statistics for the Behavioral Sciences.* Cengage Learning, 2010 (Available Titles Aplia Series). – ISBN 9780840031433

[HFC05] HANCHUAN, P. ; FUHUI, L. ; C., Ding: Feature selection based on mutual information: criteria of max-dependency, max-relevance, and min-redundancy. In: *IEEE Transactions on Pattern Analysis and Machine Intelligence* 27 (2005), S. 1226–1238

[Him78] HIMMELBLAU, D. M.: *Fault detection and diagnosis in chemical and petrochemical processes.* Elsevier Scientific Pub. Co, 1978. – ISBN 0444417478

[Hor00] HORCH, A.: *Condition Monitoring of Control Loops.* Tekniska hoegsk., 2000. – ISBN 9789171706386

[HP09] HAGENMEYER, V. ; PIECHOTTKA, U.: Innovative Prozessführung, Erfahrungen und Perspektiven. In: *atp* (2009), S. 46–63

[Hum96] HUME, D.: *A Treatise of Human Nature).* Clarendon Press, 1896

[Ing07] INGHAM, J.: *Chemical Engineering Dynamics: An Introduction to Modelling and Computer Simulation.* Wiley-VCH, 2007. – ISBN 9783527316786

[Ise06] ISERMANN, R.: *Fault-Diagnosis Systems.* Springer, 2006. – ISBN 978–3–540–24112–6

[Joa00] JOACHIMS, T.: Estimating the Generalization Performance of an SVM Efficiently. In: *Proceedings of the Seventeenth International Conference on Machine Learning,* Morgan Kaufmann Publishers Inc., 2000 (ICML '00). – ISBN 1–55860–707–2, S. 431–438

[Joh93] JOHANSSON, R.: *System modeling and identification.* Prentice Hall, 1993 (Prentice-Hall information and system sciences series)

[JRWW98] JEFFREY, C. ; REEDS, A. ; WRIGHT, M. H. ; WRIGHT, P. E.: Convergence Properties of the Nelder-Mead Simplex Method in Low Dimensions. In: *SIAM Journal of Optimization* 9 (1998), S. 112–147

[KESM02] KAUSHIK, C. ; EAMONN, K. ; SHARAD, M. ; MICHAEL, P.: Locally Adaptive Dimensionality Reduction for Indexing Large Time Series Databases. In: *Proceedings of ACM SIGMOD Conference on Management of Data,* 2002, S. 151–162

[KG97] KOHAVI, R ; GEORGE, H.: Wrappers for Feature Subset Selection. In: *Artif. Intell.* 97 (1997), Nr. 1-2, S. 273–324

[KV11] KRTEK, J. ; VOSVRDA, M.: Comparing Neural Networks and ARMA Models in Artificial Stock Market. In: *Bulletin of the Czech Econometric Society* 18 (2011), Nr. 28

[Lea88] LEAR, J.: *Aristotle: The desire to understand .* 1. Cambridge University Press, 1988. – ISBN 0521347629

[LFR06] LUND, T. ; FAULTNER, E. ; ROBINSON, M.: Condition Monitoring Using Bayesian Networks. In: *INFORMS Annual Meeting 2006* (2006)

[LKB04] LAST, M. ; KANDEL, A. ; BUNKE, H.: *Data mining in time series databases*. World Scientific, 2004 (Series in machine perception and artificial intelligence). – ISBN 9789812382900

[LM98] LIU, H. ; MOTODA, H.: *Feature Selection for Knowledge Discovery and Data Mining*. Norwell, MA, USA : Kluwer Academic Publishers, 1998. – ISBN 079238198X

[Lun10] LUNZE, J.: *Regelungstechnik 1: Systemtheoretische Grundlagen, Analyse und Entwurf einschleifiger Regelungen*. Springer Berlin Heidelberg, 2010. – ISBN 3642138071

[Man08] MANABU, K.: Data-based process monitoring, process control, and quality improvement: Recent developments and applications in steel industry. In: *Computers and Chemical Engineering* (2008)

[Mar03] MARGARITIS, D.: *Learning Bayesian Network Model Structure from Data*. Pittsburgh, PA, School of Computer Science, Carnegie-Mellon University, Diss., May 2003

[May70] MAY, W.: Knowledge of Causality in Hume and Aquinas. In: *The Thomist* (1970)

[May09] MAYRHOFER, R.: *Kausales Denken, Bayes-Netze und die Markov-Bedingung*. Universitaet Göttingen, 2009

[McK98] MCKINNON, K. I. M.: Convergence of the Nelder–Mead Simplex Method to a Nonstationary Point. In: *SIAM J. on Optimization* 9 (1998), Mai, Nr. 1, S. 148–158. – ISSN 1052–6234

[Med06] MEDER, B.: *Seeing versus doing: causal Bayes nets as psychological models of causal reasoning*. Universitaet Goettingen, 2006

[MGS05] MARKOWETZ, F. ; GROSSMANN, S. ; SPANG, R.: Probabilistic Soft Interventions in Conditional Gaussian Networks. In: *In 10th AI/Stats* (2005)

[Mik09] MIKUT, R.: *Data Mining in der Medizin und Medizintechnik.* Universitätsverlag Karlsruhe, 2009

[MK02] MARSCHINSKI, R. ; KANTZ, H.: Analysing the information flow between financial time series. In: *The European Physical Journal B – Condensed Matter* 30 (2002), Nr. 2, S. 275–281

[MR03] MEIR, R. ; RÄTSCH, G.: Advanced lectures on machine learning. (2003), S. 118–183. ISBN 3–540–00529–3

[Mur02] MURPHY, K. P.: *Dynamic Bayesian Networks: Representation, Inference and Learning,* University of California, Berkeley, Diss., 2002

[Nea09] NEAPOLITAN, R. E.: *Probabilistic Methods for Bioinformatics: with an Introduction to Bayesian Networks.* Morgan Kaufmann Publishers Inc., 2009. – ISBN 0123704766

[NM65] NELDER, J. A. ; MEAD, R.: A Simplex Method for Function Minimization. In: *The Computer Journal* (1965), S. 308–313. – ISSN 1460–2067

[OD08] OLSEN, D. ; DELEN, D.: *Advanced Data Mining Techniques.* Springer, 2008

[Pau81] PAU, L. F.: *Failure diagnosis and performance monitoring.* M. Dekker, 1981 (Control and systems theory). – ISBN 9780824710187

[PB03] PETER, F. ; BLOCKEEL, H.: *Decision support for data mining: An introduction to ROC analysis and its applications.* Kluwer, 2003

[Pea00] PEARL, J.: *Causality: Models, Reasning and Inference.* Cambridge University Press, 2000

[PG93] PINDER, A. C. ; GODFREY, G.: *Food Process Monitoring Systems.* Blackie Academic & Professional, 1993. – ISBN 9780751400991

[Pyl99] PYLE, D.: *Data Preparation for Data Mining.* Morgan Kaufmann, 1999. – ISBN 1–55860–529–0

[Rak07] RAKOTOMAMONJY, A.: Analysis of SVM regression bounds for variable ranking. In: *Neurocomputing* 70 (2007), S. 1489–1501

[Ris78] RISSANEN, J.: Paper: Modeling by shortest data description. In: *Automatica* 14 (1978), September, Nr. 5, S. 465–471. – ISSN 0005–1098

[RM06] RAGOT, J. ; MAQUIN, D.: Fault measurement detection in an urban water supply network. In: *Journal of Process Control* 16 (2006), October, Nr. 9, S. 887–902. – ISSN 09591524

[RP06] RAINER, D. ; PFEIFFER, B.: Modellbasierte prädiktive Regelung in der industriellen Praxis (Industrial Application of Model Predictive Control). In: *Automatisierungstechnik* 54 (2006), Nr. 12, S. 590–601

[Sau79] SAUNDERS, A.: The Short-Run Causal Relationship between U.K. Interest Rates, Share Prices and Dividend Yields. In: *Scottish Journal of Political Economy* 26 (1979), Nr. 1, S. 61–71

[SB11] SCUTARI, M. ; BROGINI, A.: Bayesian Network Structure Learning with Permutation Tests. In: *Statistics for Complex Problems: the Multivariate Permutation Approach and Related Topics*, 2011

[Sch00] SCHREIBER, T.: Measuring information transfer. In: *Physical review letters* 85 (2000), Nr. 2, S. 461–464

[Sch10] SCHLICHTHÄRLE, D.: *Digital Filters: Basics and Design.* Springer Verlag, 2010. – ISBN 9783642143243

[Set10] SETH, A. K.: A MATLAB toolbox for Granger causal connectivity analysis. In: *Journal of neuroscience methods* 186 (2010), Nr. 2, S. 262–273. – ISSN 1872–678X

[SF12] SCHAPIRE, R. ; FREUND, Y.: *Boosting: Foundations and Algorithms.* MIT Press, 2012

[SGS00] SPIRTES, P. ; GLYMOUR, C. ; SCHEINES, R.: *Causation, Prediction, and Search.* MIT press, 2000

[SGT$^+$09] SABESAN, S. ; GOOD, L.B. ; TSAKALIS, K.S. ; SPANIAS, A. ; TREIMAN, D.M. ; IASERMIDIS, L.D.: Information flow and application to epileptogenic focus localization from intracranial EEG. In: *IEEE Trans. Neural. Syst. Rehabil. Eng.* 3 (2009), Nr. 17, S. 244–253

[SH06] SACHS, L. ; HEDDERICH, J.: *Angewandte Statistik.* Springer-Verlag Berlin Heidelberg, 2006

[Sha48] SHANNON, C. E.: A mathematical theory of communication. In: *Bell system technical journal* 27 (1948), S. 379–423

[She00b] SHEARER, C.: The CRISP-DM Model: The New Blueprint for Data Mining. In: *Journal of Data Warehousing 5* 4 (2000), S. 13–22

[SK97] SAJIDMAN, M. ; KUNTZE, H.-B.: Integration of fuzzy control and model based concepts for disturbed industrial plants with large dead-times. In: *Fuzzy Systems, 1997., Proceedings of the Sixth IEEE International Conference on* Bd. 2, 1997, S. 1007 –1013 vol.2

[SL09] STANIEK, M. ; LEHNERTZ, K.: Symbolic transfer entropy: Inferring directionality. In: *Biomed Tech* 54 (2009), S. 323–328

[SPP$^+$05] SACHS, K. ; PEREZ, O. ; PE'ER, D. ; LAUFFENBURGER, D. A. ; NOLAN, G. P.: Causal Protein-Signaling Networks Derived from Multiparameter Single-Cell Data. In: *Science* 308 (2005), Nr. 5721, S. 523–529

[SS01] SCHÖLKOPF, B. ; SMOLA, A. J.: *Learning with Kernels: Support Vector Machines, Regularization, Optimization, and Beyond.* MIT Press, 2001. – ISBN 0262194759

[SS04]   SMOLA, A. J. ; SCHÖLKOPF, B.: A tutorial on support vector regression. In: *Statistics and Computing* (2004), S. 199–222

[SS05]   SINGHAL, A. ; SALSBURY, T. I.: A simple method for detecting valve stiction in oscillating control loops. In: *Journal of Process Control* 15 (2005), Nr. 4, S. 371 – 382

[STC04]   SHAWE-TAYLOR, J. ; CRISTIANINI, N.: *Kernel Methods for Pattern Analysis*. Cambridge University Press, 2004. – ISBN 0521813972

[TBA06]   TSAMARDINOS, I. ; BROWN, L. E. ; ALIFERIS, C. F.: The max-min hill-climbing Bayesian network structure learning algorithm. In: *Machine Learning* 65 (2006), S. 31–78

[TR02]   TENNY, M. ; RAWLING, J.: Closed-loop Behavior of Nonlinear Model Predictive Control. (2002)

[TS09]   TAFAZZOLI, E. ; SAIF, M.: Application of combined support vector machines in process fault diagnosis. In: *Proceedings of the American Control Conference*, IEEE Press, 2009 (ACC'09), S. 3429–3433

[UGP96]   USAMA, F. ; GREGORY, P. ; PADHRAIC, S.: From Data Mining to Knowledge Discovery in Databases. In: *AI Magazine* 17 (1996), S. 37–54

[Van97]   VANDERBEI, R. J.: LOQO: An interior point code for quadratic programming. In: *Optimization Methods and Software* 11 (1997), S. 451–484

[Vap82]   VAPNIK, V. N.: *Estimation of Dependences Based on Empirical Data*. Springer, 1982

[Vap95]   VAPNIK, V. N.: *The Nature of Statistical Learning Theory*. Springer, 1995

[Vap98]   VAPNIK, V. N.: *Statistical Learning Theory*. Wiley-Interscience, 1998. – ISBN 0471030031

[VC00]   VAPNIK, V. N. ; CHAPELLE, O.: Bounds on Error Expectation for Support Vector Machines. In: *Neural Computation* 12 (2000), Nr. 9, S. 2013–2036

[Vog08]   VOGT, M.: Support Vector Machines for Identification and Classification Problems in Control Engineering. In: *Automatisierungstechnik* 56 (2008), Nr. 7, S. 391–392

[WF05]   WITTEN, I. H. ; FRANK, E.: *Data Mining: Practical machine learning tools and techniques.* 2. Morgan Kaufmann, 2005

[WKRQ⁺07] WU, X. ; KUMAR, V. ; ROSS QUINLAN, J. ; GHOSH, J. ; YANG, Q. ; MOTODA, H. ; MCLACHLAN, G. J. ; NG: Top 10 algorithms in Data Mining. In: *Knowl. Inf. Syst.* 14 (2007), S. 1–37

[WMD02]   WEIDL, G. ; MADSEN, A.L. ; DAHLQUIST, E.: Condition Monitoring, Root Cause Analysis and Decision Support on Urgency of Actions. In: *Frontiers in Artificial Intelligence and Applications* 87 (2002)

[XKR⁺10]   XIAOYAN, M. ; KEWEI, C. ; RUI, L. ; XIAOTONG, W. ; LI, Y. ; XIA, W: Application of Granger causality analysis to effective connectivity of the default-mode network. In: *Complex Medical Engineering (CME), 2010 IEEE/ICME International Conference on*, 2010, S. 156 –160

[YL10]   YANG, W. ; LUO, Q.: Modeling Protein-Signaling Networks with Granger Causality Test. In: *Frontiers in Computational and Systems Biology* Bd. 15. Springer London, 2010, Kapitel 13, S. 249– 257

[YX12]   YANG, F. ; XIAO, D.: Progress in Root Cause and Fault Propagation Analysis of Large-Scale Industrial Processes. In: *Journal of Control Science and Engineering* (2012)

[ZYSM07]   ZEXUAN, Z. ; YEW-SOON, O. ; MANORANJAN, D.: Wrapper-Filter Feature Selection Algorithm Using a Memetic Framework. In:

IEEE Transactions on Systems, Man, and Cybernetics, Part B
(2007), Nr. 1, S. 70–76

Own Publications

[KB09a] KÜHNERT, C. ; BERNARD, T.: Extraction of Optimal Control Patterns in Industrial Batch Processes based on Support Vector Machines. In: *Control Applications, (CCA) & Intelligent Control (ISIC), IEEE* Bd. 1, 2009. – ISBN 978–1–4244–4602–5, S. 481–486

[KB09b] KÜHNERT, C. ; BERNARD, T.: Optimization and Online-Monitoring in Industrial Batch Processes using Data Mining Methods. In: *19. Workshop Computational Intelligence*, KIT Scientific Publishing, 2009. – ISBN 978–3–86644–434–8, S. 170–180

[KB10a] KÜHNERT, C. ; BERNARD, T.: Online-Monitoring und Prädiktion der Prozessgüte komplexer Batchprozesse. In: *Expertenforum Verteilte Messsysteme*, KIT Scientific Publishing, 2010, S. 13–24

[KB10b] KÜHNERT, C. ; BERNARD, T.: Online-Monitoring und Prädiktion der Prozessgüte komplexer Batchprozesse mittels informationstheoretischer Maße und Support Vector Machines. In: *Automatisierungskongress 2010*, VDI-Verlag, 2010, S. 107–111

[KBF10] KÜHNERT, C. ; BERNARD, T. ; FREY, C.: Erlernen kausaler Zusammenhänge aus Messdaten mittels gerichteter azyklischer Graphen. In: *20. Workshop Computational Intelligence*, KIT Scientific Publishing, 2010, S. 172–188. – Young Author Award Runner-Up

[KBF11] KÜHNERT, C. ; BERNARD, T. ; FREY, C. W.: Causal Structure Learning in Process Engineering Using Bayes Nets and Soft Interventions. In:

International Conference on Industrial Informatics, IEEE, 2011, S. 69–74

[KGHM11] KÜHNERT, C. ; GRÖLL, L. ; HEIZMANN, M. ; MIKUT, R.: Ansätze zur datengetriebenen Formulierung von Strukturhypothesen für dynamische Systeme. In: *21. Workshop Computational Intelligence,* KIT Scientific Publishing, 2011, S. 15–30

[KGHM12] KÜHNERT, C. ; GRÖLL, L. ; HEIZMANN, M. ; MIKUT, R.: Methoden zur datengetriebenen Formulierung und Visualisierung von Kausalitätshypothesen. In: *at-Automatisierungstechnik* 60 (2012), Nr. 10, S. 630–640

[KMB09] KÜHNERT, C. ; MINX, J. ; BERNARD, T.: Optimierung der Prozessführung komplexer verfahrenstechnischer Prozesse mit Support Vector Machines. In: *Automatisierungskongress 2009,* VDI-Verlag, 2009. – ISBN 978–3–18–092067–2, S. 233–236

Karlsruher Schriftenreihe zur Anthropomatik
(ISSN 1863-6489)

Herausgeber: Prof. Dr.-Ing. Jürgen Beyerer

Die Bände sind unter www.ksp.kit.edu als PDF frei verfügbar
oder als Druckausgabe bestellbar.

Band 1   Jürgen Geisler
Leistung des Menschen am Bildschirmarbeitsplatz. 2006
ISBN 3-86644-070-7

Band 2   Elisabeth Peinsipp-Byma
**Leistungserhöhung durch Assistenz in interaktiven Systemen
zur Szenenanalyse.** 2007
ISBN 978-3-86644-149-1

Band 3   Jürgen Geisler, Jürgen Beyerer (Hrsg.)
Mensch-Maschine-Systeme. 2010
ISBN 978-3-86644-457-7

Band 4   Jürgen Beyerer, Marco Huber (Hrsg.)
**Proceedings of the 2009 Joint Workshop of Fraunhofer IOSB and
Institute for Anthropomatics, Vision and Fusion Laboratory.** 2010
ISBN 978-3-86644-469-0

Band 5   Thomas Usländer
Service-oriented design of environmental information systems. 2010
ISBN 978-3-86644-499-7

Band 6   Giulio Milighetti
**Multisensorielle diskret-kontinuierliche Überwachung und
Regelung humanoider Roboter.** 2010
ISBN 978-3-86644-568-0

Band 7   Jürgen Beyerer, Marco Huber (Hrsg.)
**Proceedings of the 2010 Joint Workshop of Fraunhofer IOSB and
Institute for Anthropomatics, Vision and Fusion Laboratory.** 2011
ISBN 978-3-86644-609-0

Band 8   Eduardo Monari
**Dynamische Sensorselektion zur auftragsorientierten
Objektverfolgung in Kameranetzwerken.** 2011
ISBN 978-3-86644-729-5

Band 9   Thomas Bader
Multimodale Interaktion in Multi-Display-Umgebungen. 2011
ISBN 3-86644-760-8

Band 10   Christian Frese
Planung kooperativer Fahrmanöver für kognitive Automobile. 2012
ISBN 978-3-86644-798-1

Band 11   Jürgen Beyerer, Alexey Pak (Hrsg.)
**Proceedings of the 2011 Joint Workshop of Fraunhofer IOSB and
Institute for Anthropomatics, Vision and Fusion Laboratory.** 2012
ISBN 978-3-86644-855-1

Band 12   Miriam Schleipen
**Adaptivität und Interoperabilität von Manufacturing Execution
Systemen (MES).** 2013
ISBN 978-3-86644-955-8

Band 13   Jürgen Beyerer, Alexey Pak (Hrsg.)
**Proceedings of the 2012 Joint Workshop of Fraunhofer IOSB and
Institute for Anthropomatics, Vision and Fusion Laboratory.** 2013
ISBN 978-3-86644-988-6

Band 14   Hauke-Hendrik Vagts
**Privatheit und Datenschutz in der intelligenten Überwachung:
Ein datenschutzgewährendes System, entworfen nach dem
„Privacy by Design" Prinzip.** 2013
ISBN 978-3-7315-0041-4

Band 15   Christian Kühnert
**Data-driven Methods for Fault
Localization in Process Technology.** 2013
ISBN 978-3-7315-0098-8